High Resolution Fourier Transform Spectroscopy of Linear Molecules

HIGH RESOLUTION FOURIER TRANSFORM SPECTROSCOPY OF LINEAR MOLECULES

K. A. MOHAMED

Nova Science Publishers, Inc.
New York

For permission to use material from this book please contact us:
Telephone 631-231-7269; Fax 631-231-8175
Web Site: http://www.novapublishers.com

Library of Congress Cataloging-in-Publication Data
Available upon request

ISBN 1-59454-171-X

Published by Nova Science Publishers, Inc. ✦ New York

CONTENTS

PREFACE

High resolution Fourier transform spectra of linear molecules have evoked a great deal of interest during the last several years, which could be seen from the several hundreds of papers published in the Journals on the spectra of diatomic and small linear polyatomic molecules. This review describes the advantages of FT spectroscopy, the techniques employed in absorption and emission spectroscopy, and presents the theoretical models and formulas used in the analyses and interpretations of the spectra. The perturbations observed in the spectra due to Fermi, Darling–Dennison, Coriolis, and other anharmonic resonances; and vibrational and rotational *l*–type resonances etc. are discussed with suitable examples. The types of information obtained from FT spectroscopy of two to ten atomic linear molecules, their observed transitions and spectral perturbations etc. are presented alongwith some figures and curves, reproduced from the published literature. In the analyses of the data, several parameters are employed in the diagonal and off–diagonal matrix elements of the Hamiltonian. Some molecular parameters for few selected linear molecules are reproduced from published work, to serve as examples for the theoretical formulations and global analyses. Since Fourier transform spectroscopic work is still a very active field of molecular research, as could be discerned from the current literature, a comprehensive review on the topic is considered to be both timely and appropriate.

Chapter 1

INTRODUCTION

The technique of Fourier transform spectroscopy enjoys superiority over prism and grating spectrographic techniques and also over other techniques like laser spectroscopy. The fundamental advantage of an interferometer over a scanning monochromator is due to the Fellgett's advantage or the multiplex advantage according to which, an interferometer gathers information at all frequencies simultaneously. The second important advantage is the Jacquinot's advantage due to which, at a given resolution, more light could be accepted in an interferometer. A third advantage is the connes' advantage or the accuracy of frequency determination, made possible by the use of a helium-neon laser to reference the moving mirror in the interferometer. The FT spectrometer has also very important data processing advantages with the incorporation of a minicomputer dedicated to the instrument. In additon, large Fourier transform computations have been made easy with the invention of the Fast Fourier Transform (FFT) algorithm by Cooley and Tukey, which reduced the number of steps for a transform of N samples from N^2 to $N\log_2 N$. Thus with very large resolving power, high wavenumber accuracy, reduced stray light problems and fast scanning and computing times, FT spectroscopic technique has become one of the most powerful tool to record and analyze high resolution molecular spectra. The technique is particularly more important in the near and mid infrared regions where high resolution is of immense experimental importance.

The principles, techniques and advantages of FTS have been described in books [1-3] and review articles [4-7]. The excellent volumes on the applications of FT spectroscopy in various fields edited by Ferraro and Basile [8-11] provide a wealth of knowledge on the topic. The technique has been applied to study the spectra of hundreds of molecules of various shapes, masses and atomic compositions during the last several years. Linear molecules containing two or more atoms in gaseous phase form the largest set of molecules investigated. Light molecules and their isotopomers possess large rotational constants and hence could be studied in finest detail.

The analysis of spectral perturbations by FTS technique is an important contribution. Perturbations cause shift in energy levels and hence in the line positions of the vibrational-rotational transitions. There are two types of perturbations viz, Intramolecular Vibrational Relaxation (IVR) and Intramolecular Vibration Rotation Energy Transfer (IVRET). IVR is the process by which vibrational energy is exchanged amongst the various vibrational modes of a molecule. The mechanisms of IVR are vibrational resonances of the Fermi type [12] and of the Darling-Dennison type [13]. The IVRET processes show the existence of rotational motion mixing the vibrational states. Coriolis coupling and interactions are examples of IVRET processes. In addition to the types of interactions causing IVR and IVRET, there are perturbations in the form of the vibrational and rotational *l*-type resonances. FT Spectroscopy has been an ideal technique to study the spectra under high resolution involving the above mentioned interactions and resonances. Such processes will be discussed in this review.

A variety of topics in the field of molecular spectroscopy are being investigated by the FTS technique. In a nutshell, some recent work with representative examples are on the following:

- Infrared emission at high temperature, of molecules like HCN [14], HNC [15] whose ground state is about 3800 cm^{-1} above the ground state of HCN, DCN [16], of hitherto unknown free radicals like BiNa [17] and TeLi [18] and of fine structure transitions like $X_2\ ^2\Pi_{3/2} \rightarrow X_1\ ^2\Pi_{1/2}$ of PbF and PbCl [19].
- Analysis of Rydberg-Rydberg transitions like in ArH and ArD [20].
- Use of Laser Induced Fluorescence (LIF) coupled with FTS to analyze decay radiation [21-23] so that the advantages of both techniques could be exploited.
- Determination of potential energy curves from experimentally obtained molecular parameters e.g. for Br_2 [23], and theoretical verification of experimentally obtained potential curves, like for Rb_2 [24].
- Absolute line intensity measurements [25-28].
- Width and shifts of lines [29, 30], asymmetry of line profiles [31], and broadening coefficients, e.g. for NO fundamental vibrational band by different noble gases as broadening species [32] and by O_2 [33].
- Collision induced absorption spectroscopy, like in CO_2 [34]. The strong Fermi coupled doublet (ν_1, $2\nu_2$) in CO_2 is forbidden in IR in absorption by the principle of mutual exclusion. But in pressurized CO_2, the doublet has been observed.
- Simultaneous or double vibrational collision induced absorption in gas mixtures, like in the process CO_2 (v_3=1) + N_2 (v=1) ← CO_2 (v_3=0) + N_2 (v=0) [35].
- Overtone spectroscopy [36].
- Vibrational transition moments [37].
- Magnetic dipole transitions [38].
- Systematic study of absorption bands close to dissociation limit [39], band oscillator strengths [40], absolute cross sections for collision induced bands [41] and improved data set [42] for the same band system, like the Herzberg band system of O_2 [39-42].
- Spectra and integrated band intensities of long carbon chain linear molecules (polyynes) like C_8H_2 [43] and C_6H_2 [44], which give parameters to constrain planetary atmosphere modelling.

The coming sections of the present review deal with some experimental aspects of FTS, necessary theoretical considerations, and discuss the types of information obtained from FT spectra; and present data on molecular parameters for selected diatomic, triatomic, tetraatomic and other polyatomic linear molecules. The molecules and the papers cited in this review are chosen only as a representative of the type of results and information obtained from FTS. In this process, many excellent works on the subject have been omitted from consideration by the necessity of limiting the size of the review.

Chapter 2

EXPERIMENTAL ASPECTS OF HIGH RESOLUTION FOURIER TRANSFORM SPECTROSCOPY

In this section, a brief description of the types of FT Spectrometers, and some experimental techniques used in high resolution absorption and emission spectroscopy are given. The calibration of spectra are also discussed.

2.1 FOURIER TRANSFORM SPECTROMETERS

There are several commercial FT spectrometers being used in the high resolution spectroscopy of linear molecules. The design of most instruments is based on the two beam interferometer originally designed by Michelson in 1891. The technical developments are the use of a dedicated minicomputer and a helium-neon laser to monitor the travel of the moving mirror. The use of the laser allows the interferogram to be digitized at precisely equal intervals for purposes of coaddition of scans to improve signal-to-noise ratio, and also provides an internal wavenumber standard. A high speed vector processor system to assist the computer is an additional feature in some commercial instruments.

The apodized resolution of an FT interferometer depends on the reciprocal of the maximum optical path difference (MOPD). Some workers have modified commercial FT spectrometers to enhance their performance or fabricated new ones for more accurate results. Plummer et al [45] have modified a Bruker IFS 120 HR spectrometer so that its MOPD was made 4 metres and the maximum apodized resolution obtained was 0.0025 cm^{-1}. A modified Bomem DA 3.002 FTIR spectrometer with MOPD of 4.45 metres and having an unapodized resolution of 0.0014 cm^{-1} has been used by Kabbadj et al [46] and by Di Lonardo et al [47]. Valentin [48] has reported the fabricational and technical aspects of a 22 metre maximum path difference FT spectrometer. A folded Michelson interferometer with cats eye retroreflectors has been described by Plateaux et al [49] who have achieved effective relative wavelength stabilization of better than 10^{-10} by using a He-Ne laser discharge tube alongwith a temperature regulated intracavity iodine absorption cell.

2.2 High Resolution Absorption Spectroscopy

For absorption studies, the sample in gaseous form or in vapour state is irradiated with the output radiation from the FT spectrometer. The transmitted light intensity I (x) plotted against the optical path difference x constitutes the interferogram. The optical path difference is varied slowly by displacing the moving mirror at constant velocity. The modulation frequency of the interferogram (Fourier frequency) is $2v\tilde{\nu}$, where v is the velocity of the moving mirror and $\tilde{\nu}$ is the wavenumber content of the incident radiation. In absorption work, one usually estimates the wavenumber dependent transmittance T ($\tilde{\nu}$) determined as the ratio of the spectral intensity with the sample at wavenumber $\tilde{\nu}$ to the spectral intensity without sample at $\tilde{\nu}$. Sometimes the absorbance $A(\tilde{\nu}) = -\log_{10} T(\tilde{\nu})$ is reported.

The preparation/synthesis of the gaseous sample under study and the use of the absorption cells are different in different cases. Single-pass gas cells are sufficient to get the desired strength of absorption signal in many cases. Thus, in the recording of the FTIR spectrum of diacetylene (C_4H_2), Khlifi et al [50] used a 10 cm long gas cell. The high resolution spectrum of triacetylene (C_6H_2) in the 27-227 cm^{-1} region (ν_{13} fundamental band) was reported at a resolution of 0.0018 cm^{-1} by Haas et al [51] by using a 3 metre long single pass absorption cell. A 4 metre cell to record the FTIR spectrum of BrCN [52] in the 395-765 cm^{-1} region, and the use of 28 cm and 20 cm long glass cells to record the spectrum of DCCCl [53] in the ν_1 and ν_2 band regions respectively are mentioned in recently reported work.

To record weak and hot bands, longer absorption paths are necessary. For this purpose, multiple traversal cells are used. After the synthesis of tetraacetylene (C_8H_2), Shindo et al [43] mixed it with a solvent like tetrabutyl tin $(C_4H_9)_4Sn$ to lower the vapour pressure. The vapourization of the mixture was controlled by dipping the mixture in a cold bath at –20°C. A multiple reflection cell was then employed with path length fixed at 10.6m to record the FTIR spectrum. White type [54] multiple reflection cells of various base lengths and total path lengths have been employed by various workers. Table 2.1 shows some examples of the types of White cells, total path lengths and gas pressures used to study the FT absorption spectra of few linear molecules.

Table 2.1. Types of White cells, total absorption path lengths and gas pressures employed in the high resolution FT spectral recording of few linear molecules.

Molecule	Pressure (Torr)	White cell base length (m)	Total path length employed (m)	Reference
C_2H_2	0.75	0.50	18	[46]
C_2H_2	7.7	0.156	1.872	[25]
C_2H_2	0.03-3.0	-	32	[55]
CO_2	5.75	-	40.18	[26]
CO_2	60.1	25	1507	[56]
FCN	0.023-3.65	2	64-104	[57]
LiF, LiCl	100	2	32	[58]
N_2O	0.74-1.02	6	25-433	[59]
N_2O	0.77-27.6	1.72	6.98-48.26	[27]
O_2	20-750	50	201.84,402.08,602.32	[40,41]
^{18}OCS	0.0075-0.60	2	64	[60]

2.3 HIGH RESOLUTION EMISSION SPECTROSCOPY

The emission spectra of very large number of diatomic and few small linear triatomic molecules in the UV, visible and near IR regions using the FTS technique have been reported during the last few years. The multiplex advantage associated with the FT technique is the prime reason for the large research output in this area.

The emission spectra have been produced by different techniques with certain features, some of which will be reviewed here. The technique of laser induced fluorescence (LIF) coupled with FT spectroscopy yields the advantages of both techniques. Fellows et al [61] produced the heaviest heteronuclear molecule RbCs in a heat pipe oven and then excited the molecule by a Ti: Sa laser pumped by an Ar^+ laser. By exciting the $A^1\Sigma^+$ electronic state of RbCs, it was possible to obtain fluorescence upto the highest vibrational levels of the ground state $X^1\Sigma^+$, due to the relative positions of the internuclear distances between the $A^1\Sigma^+$ and $X^1\Sigma^+$ states. The fluorescence between $A^1\Sigma^+$ and $X^1\Sigma^+$ states was analyzed by an FT spectrometer. The ranges of observed v″ and J″ were 119 and 250 respectively for RbCs in the experiment of Fellows et al [61]. The LIF+FTS techniques have been applied to study the emission spectra of molecules like Br_2 [23], $CuCl_2$ [62], KRb [63], NiCl [64] and BaI [21]. In the case of KRb, the molecule was produced in a steel pipe oven by heating K and Rb metals. Excitation process $(3)^1\Pi \leftarrow X^1\Sigma^+$ was achieved by fixed frequency Ar^+ laser lines and the fluorescence due to the $(3)^1\Pi \rightarrow (2)^3\Sigma^+$ transition was recorded on an FT spectrometer by Amiot [63]. The $(3)^1\Pi$ state is strongly perturbed by the $(4)^3\Pi$ state. Transition from this mixed singlet-triplet character upperstate down to $(2)^3\Sigma^+$ gave the fluorescence [63].

High temperature thermal emission from molecules like HCN [14], HNC [15], DCN [16], BaI [21], LaF [65] and GeO [66] have been focused onto FT spectrometers and spectra analyzed. In the case of GeO [66], the emission was observed by heating Ge in a furnace upto a temperature of 1500°C and by passing 2 Torr oxygen and 18 Torr argon as buffer gas in the observation cell.

Radiation due to chemiluminiscence has been used to obtain emission spectra of several diatomic molecules like SO [38], BiO [67], SrO [68] and BaI [69]. Setzer et al [38] observed chemiluminiscence spectum of SO in a fast flow system. SO molecules were generated in a side arm of an observation cell by passing a microwave discharged mixture of oxygen and argon carrier gas over molten sulphur. Metastable oxygen molecules O_2 $(a^1\Delta_g)$ were generated in another side arm of the observation cell by passing O_2 through a second microwave discharge and over mercury oxide. The flow system consisted of a glass tube observation cell with quartz end windows to which the two side arms were attached. The SO molecules were excited to their $a^1\Delta$ and $b^1\Sigma^+$ states by electronic-to-electronic exchanges from O_2 $(a\ ^1\Delta_g)$ molecules and by energy pooling processes. The emission from the $b^1\Sigma^+ \rightarrow X^3\Sigma^-$ and $a^1\Delta \rightarrow X^3\Sigma^-$ transitions of SO was then analyzed by FT spectroscopy by Setzer et al [38].

SO is isoelectronic with O_2 and thus has a $X^3\sum^-$ ground state and low lying metastable $a^1\Delta$ and $b^1\sum^+$ excited states. Consequently, the $a^1\Delta \rightarrow X^3\sum^-$ transition is 'forbidden'.

Microwave discharges through flowing mixture of gases/vapours have been used to excite and study FT emission spectra of molecules like TiCl [70], NiCl [64], InCl [71], SeH [72], BiNa [17] and NbN [73]. Quite recently, Ram et al [74] have studied the emission spectrum of VN. The bands of VN were produced in a cell by the reaction of flowing $VOCl_3$ vapour and active nitrogen. Active nitrogen was produced in a short quartz tube attached to the reaction cell by flowing N_2 through a microwave discharge.

Other methods to excite spectra reported in the field of FT emission spectroscopy include AC, DC and hollow cathode discharges through gases. High voltage AC discharge at 70 KHz and discharge currents upto 700 mA and 200 mA respectively were used by Imajo et al to record emission from PN^+ ions [75] and from TiCl [76]. DC discharges were used in the study of the spectra of NeH^+ [37] and CO_2 [77]. Campargue et al [77] observed that it was possible to populate very high vibrational states v_3 of CO_2 when excited by a DC discharge in a $CO_2/N_2/He$ mixture at typically 7KV, 60 mA. Helium efficiently decreases kinetic temperature and when added to CO_2+N_2 mixture, population of v_3 upto 6 were observed. Hollow cathode discharge lamps have been used to study FT emission spectra of molecules like BeD [78] and CuCl [79].

There are other methods reported like radiofrequency discharge at 13.5 MHz employed by Bailly and Vervloet [80] to study the emission spectrum of N_2O. The authors noticed that the addition of N_2 with N_2O helped in increased emission from N_2O. This is an interesting observation because the energy difference between N_2 (v=1) and N_2O (00^01) levels is 107 cm^{-1} and hence a resonant vibrational energy transfer (like in CO_2+N_2 mixture) is not very probable.

As an example of FT studies of flame spectra, it is worth pointing out the work of Bailly et al [81] who have reported the emission spectrum in the 3μm region of CO_2, produced in a CH_4+O_2 flame.

2.4 Calibration of High Resolution FT Spectra

The high resolution molecular spectra recorded over FT spectrometers yield very improved quality in the measurement of line positions and line profiles. The wavenumbers of ro-vibrational lines could be determined with high precision.

Commercial and fabricated FT spectrometers are incorporated with a He-Ne laser which provides internal wavenumber calibration with high accuracy. Using this internal wavenumber calibration, O'brien et al [82] have reported the near infrared $Y^2\sum^+ \rightarrow X^2\Pi$ transition of CuS in emission (10,000-11,000 cm^{-1} region) and estimated the absolute accuracy of wavenumber measurement to be better than 0.004 cm^{-1}.

The absolute calibration of spectra are made by using the very accurate wavenumbers of other molecules and atoms. These molecules may occur as impurity (like H_2O) inside the absorption cell or added alongwith the sample to obtain their characteristic ro-vibrational lines spread over the spectral region of interest. In general, ro-vibrational lines of molecules like CO, CO_2, C_2H_2, CH_4, C_2H_4, CH_3Cl, CH_3I, CS_2, H_2, HCl, HF, H_2O, I_2, NO, N_2O, NH_3 and

OCS, and atomic lines of Hg, Fe I and Fe II have been used as standards. The wavenumbers of several ro-vibrational lines to be used as standards between 3 and 2600 μm have been given in the handbook of Guelachvili and Narahari Rao [83]. Recommendations of the International union of pure and applied chemistry (IUPAC) by the 'Commission on molecular structure and spectrocopy' entitled "High Resolution Wavenumber Standards for the Infrared" are also available [84] which gives the list of reference molecules, their spectral ranges and other details. In some cases, due to practical reasons, secondary standards are used, an example being the calibration of 3μm spectral region of $^{13}C_2H_2$ by Alboni et al [85].

Chapter 3

THEORETICAL CONSIDERATIONS

In this section, certain necessary formulas pertaining to the description of energy levels and vibration–rotation spectra of linear molecules are presented. These include expressions for energy, intensity, anharmonic resonances and interactions etc. Molecular constants and interaction parameters form the bulk of the data reported in the literature, in addition to the spectra, energy levels, line positions and intensities. Some molecular constants and parameters, which are introduced in the equations of this section, and obtained from high resolution FT spectroscopy, will be presented in section 4 of this review.

3.1 DIATOMIC MOLECULES

A large number of research papers published by the FTS techniques for diatomic molecules are the emission spectra in which, mostly vibrational–rotational transitions between two electronic states are involved. There are also a number of papers reported in FT absorption and few in FT emission of molecules in which, the vibrational–rotational levels of the ground electronic state are involved.

As in the case of linear polyatomic molecules, the rotationless *e/f* parity labeling of rotational levels is employed in diatomic molecules also. Rotational levels with $+(-1)^J$ are labeled *e* and those with $-(-1)^J$ are labeled *f* for integer J, while for half integer J, levels with $+(-1)^{J-1/2}$ are labeled *e* and those with $-(-1)^{J-1/2}$ are labeled *f* levels [86]. The allowed transitions are

$$e \leftrightarrow f \text{ for } \Delta J=0, \text{ and } e \leftrightarrow e,\ f \leftrightarrow f \text{ for } \Delta J = \pm 1 \qquad (3.1)$$

This classification applies mainly to rotational levels of Π, Δ, Φ... electronic states (Π,Δ... vibrational species for linear polyatomic molecules) which show Λ–type doubling (*l*–type doubling for linear polyatomic molecules). For molecules obeying a Morse potential function, the customary energy level expressions [ref. 87, pages, 106–107, 149] are sufficient.

The vibrational term values are given by

$$G(v) = \omega_e (v+1/2) - \omega_e x_e (v+1/2)^2 + \omega_e y_e (v+1/2)^3 + \omega_e z_e (v+1/2)^4 + \ldots \qquad (3.2)$$

The rotational term values are

$$F_v(J) = B_vJ(J+1) - D_v[J(J+1)]^2 + H_v[J(J+1)]^3 + \ldots \tag{3.3}$$

The vibrational dependence of rotational constant and the centrifugal distortion constant are given by

$$B_v = B_e - \alpha_e(v+1/2) + \gamma_e(v+1/2)^3 + \ldots \tag{3.4}$$

$$D_v = D_e + \beta_e(v+1/2) + \ldots \tag{3.5}$$

Where the symbols have their usual spectroscopic meaning.

For rotational levels of $^1\Pi$ electronic states, an additional term $\pm\{qJ(J+1) + q_D[J(J+1)]^2 + q_H[J(J+1)]^3\}$ has to be used with eq (3.3) to account for the Λ–doubling in Π states (upper sign for *e* levels and lower sign for *f* levels). q, q_D and q_H are Λ–doubling constants.

Several authors have reported the value of centrigugal distortion constant using the Kratzer relation

$$D_e = 4B_e^3/\omega_e^2 \tag{3.6}$$

and the value of α_e using the Pekeris relation

$$\alpha_e = [6(\omega_e x_e B_e^3)^{1/2}]/\omega_e - 6B_e^2/\omega_e \tag{3.7}$$

To treat the experimental data in a more refined way, additional terms are included in the energy expression. For example, Focsa et al [78] in their FT emission analysis of the $A^2\Pi - X^2\Sigma^+$ system of BeD, have used an additional term $L_v[J(J+1)]^4$ to eq (3.3) for the rotational energy of the $X^2\Sigma^+$ ground state. For the $A^2\Pi$ state, in place of $\omega_e z_e(v+1/2)^4$ of eq (3.2), Focsa et al [78] have used $\omega_e z_{1e}(v+1/2)^4 + \omega_e z_{2e}(v+1/2)^5$, and in place of $\gamma_e(v+1/2)^2$ of eq (3.4), used $\gamma_{1e}(v+1/2)^2 + \gamma_{2e}(v+1/2)^3 + \gamma_{3e}(v+1/2)^4$.

3.1.1 Theoretical Models

To analyze the experimental data on the line positions of ro–vibrational lines, a suitable theoretical model and least squares treatment of data are required. In the absence of external electric or magnetic field, the Hamiltonian is given by

$$H = H_0 + H_{rot} + H_{fs} + H_{hfs} \tag{3.8}$$

Where H_0 gives the rotationless vibronic energy T_v for the vibrational states of different electronic states. H_{rot} describes the rotational Hamiltonian given by

$$H_{rot} = B(r)\,\mathbf{R}^2 = \frac{h}{8\pi^2 \mu r^2}\,(\mathbf{J}-\mathbf{L}-\mathbf{S})^2 \tag{3.9}$$

Where B(r) is the radial part of the rotational energy operator and μ is the reduced mass of the molecule. The effect of rotational stretching is represented by replacing H_{rot} by an effective rotational Hamiltonian [88] given in Hund's coupling case (a) basis set as

$$H_{rot}^{eff} = B_v\,\mathbf{R}^2 - D_v\mathbf{R}^4 + H_v\,\mathbf{R}^6 + L_v\mathbf{R}^8 + M_v\mathbf{R}^{10} \tag{3.10}$$

where rotational constants B_v, D_v, H_v, L_v, M_v, etc. are associated with vibrational level v. If the basis set is chosen to be Hund's case (b) wavefunctions, then $\mathbf{R}^2$ may be replaced with $N(N+1)-\Lambda^2$, as given by Zare et al [88].

The quantity H_{fs} of eq (3.8) accounts for the interaction between the orbital and spin motions of the electrons, such that

$$H_{fs} = H_{so} + H_{ss} + H_{sr} \tag{3.11}$$

The spin–orbit Hamiltonian H_{so} has the form

$$H_{so} = A(r)\,\mathbf{L.S} \tag{3.12}$$

where A(r) is the spin–orbit coupling constant. When A is large (A>>BJ), it is referred to as Hund's case (a) coupling and when A is small (A<< BJ), it is referred to as Hund's case (b) coupling.

The spin–spin interaction term H_{ss} is zero for Hund's case (a) states $^2\Sigma$ and $^2\Pi$. However, it has an important contribution in $^3\Pi$ and $^3\Delta$ states.

The spin–rotation Hamiltonian H_{sr} is given by

$$H_{sr} = \gamma(r)\,\mathbf{R.S} = \gamma(r)\,(\mathbf{J}-\mathbf{L}-\mathbf{S}).\mathbf{S} \tag{3.13}$$

where γ(r) is the spin rotation constant and **R** is the rotational angular momentum operator of the nuclei, **L** is the total electronic angular momentum operator, **S** is the total electronic spin angular momentum operator, and **J** is the total angular momentum operator.

Under zero field conditions, the hyperfine structure part of Hamiltonian H_{hfs} may be disregarded.

The essential spectroscopic problem is handled by selection of an appropriate Hamiltonian and selection of a basis set (e.g., Hund's case (a) or (b)) of wavefunctions for each state, followed by numerical diagonalization of the Hamiltonian matrix to obtain the wavefunctions and energy levels. A least squares fit of the observed line positions to the calculated line positions constructed from appropriate energy levels found by diagonalizing the secular determinants of the upper and lower levels is performed to yield the molecular constants. For more accurate description of molecular energy levels, addition of more parameters are required in the energy expressions and in the Hamiltonian matrix elements.

The Hamiltonian matrix elements are different for different basis sets. As an example, the explicit matrix elements of $^2\Pi$ electronic state in Hund's case (a) basis set and using the "Unique perturber approximation" have been listed by Amiot et al [89] along with the matrix elements for the same state using the "effective Hamiltonian" method of Brown et al [90]. In their paper on the FT emission spectroscopy of OD, Amiot et al [89] have also listed the matrix elements of the $^2\Sigma^+$ state. A more correct listing of the matrix elements of the $^2\Sigma^+$ state in case (a) parity basis set has been provided by Douay et al [91], who have analyzed the FT emission spectrum of the Rydberg molecule XeH. The matrix elements for term value T, spin–orbit coupling constants A, A_D and A_H; rotational constants B, D, H and L; Λ–doubling constants p, p_D and p_H; Λ–doubling series q, q_D and q_H; and spin–rotation constants γ, γ_D and γ_H are available in the papers of Amiot et al [89] and Douay et al [91].

The spin–rotation coupling constant γ and the centrifugal correction A_D to the spin–orbit coupling constant A for doublet states in diatomic molecules have been discussed in detail by Brown and Watson [92]. The definition of a number of symbols and matrix elements in parity case (a) basis set are given by Coxon [93].

A large number of diatomic emission spectra analyzed by FT spectroscopy belong to Hund's case (b) coupling. For example, Gutterres et al [21], while analyzing the spectra of BaI by the LIF+FTS technique, have described the term values of $^2\Sigma^+$ electronic state by standard Hund's case (b) formulae

$$T = T_v + B_v N(N+1) - D_v[N(N+1)]^2 + H_v [N(N+1)]^3$$

$$+ \ldots\ldots\ldots + 1/2\ \gamma N \text{ for } e\text{–labeled levels and} - 1/2\ \gamma(N+1) \text{ for}$$

$$f\text{–labeled levels} \quad (3.14)$$

with

$$T_v = T_e + G(v) \text{ [given in eq (3.2)]} \quad (3.15)$$

$$B_v = B_e - \alpha_B (v+1/2) + \beta_B (v+ 1/2)^2 + \ldots\ldots\ldots \quad (3.16)$$

$$D_v = D_e + \alpha_D (v+ 1/2) + \ldots\ldots\ldots\ldots \quad (3.17)$$

$$\gamma = \gamma_e + \gamma_v (v+ 1/2) + \gamma_D N(N+1) + \ldots\ldots\ldots\ldots \quad (3.18)$$

where N is the total angular momentum exclusive of electron spin. The Hamiltonian matrix elements for isolated $^2\Pi$ electronic states, used by Gutterres et al [21] (for $A^2\Pi$ and $C\ ^2\Pi$ electronic states of BaI) are given in the following table.

	$^2\Pi_{3/2}$	$^2\Pi_{1/2}$
$^2\Pi_{3/2}$	$T + A/2 + (B+A_J)(X-1) - D[(X-1)^2 + X]$ $+H[(X-1)^3+X(3X-1)]$ $+(A_{JJ}/2)[3(X-1)^2+X]$ $+(q/2)\,X$	$-BX^{1/2} + 2DX^{3/2} - HX^{1/2}(3x^2+X+1)$ $+A_{JJ}X^{1/2}$ $+(q/2)\{[X^{1/2}[-1 \pm (X+1)^{1/2}]\}$ $-(p/4)\,X^{1/2}$
$^2\Pi_{1/2}$	Sym	$T - A/2 + (B+A_J)(X-1) - D[(X+1)^2+X]$ $+H[(X+1)^3 + X(3X+1)]$ $-(A_{JJ}/2)[3(X+1)^2 + X]$ $+(q/2)[X+2 \mp 2(X+1)^{1/2}]$ $+(p/2)[1 \mp (X+1)^{1/2}]$

Matrix elements were calculated using an *e*/*f* parity basis and written *e* over *f*. in the above table, and $X = (J+1/2)^2 - 1$.

Gutterres et al [21, 94] have taken into account, the vibrational dependence of the parameters by a Dunham–type variation with T_v given by eq. (3.15), B_v given by eq (3.16), D_v given by eq (3.17), and

$$A = A_e + A_v(v+1/2) + A_{vv}(v+1/2)^2 + \ldots\ldots\ldots\ldots\quad (3.19)$$

$$p = p_e + p_v(v+1/2) + p_J\,J(J+1) + \ldots\ldots + p_{vv}(v+1/2)^2 + \ldots\ldots\quad (3.20)$$

As an example in which the spin–spin interaction Hamiltonian H_{ss} is employed, the case of $^3\Phi \rightarrow (1)^3\Delta$ transition of LaF reported by Bernard et al [95] is one which the triplet models of rotational Hamiltonian in Hund's case (a) basis are used. Bernard et al [95] treated the $^3\Phi$ substates as separate singlets, and for the $^3\Delta$ state, the effective rotational elements used for the 3 x 3 (*e* or *f*) Hamiltonian energy matrix were

$$\langle ^3\Delta_1|^3\Delta_1\rangle = T - 2A + \varepsilon + \varepsilon_D X - \gamma + (B - 2A_D)X - D(X^2+2X-4).$$

$$\langle ^3\Delta_2|^3\Delta_2\rangle = T - 2(\varepsilon + \varepsilon_D X) - 2\gamma + B(X-2) - D(X^2-12).$$

$$\langle ^3\Delta_3|^3\Delta_3\rangle = T + 2A + \varepsilon + \varepsilon_D X - \gamma + (B+2A_D)(X-8) - D(X^2-14X+52).$$

$$\langle ^3\Delta_1|^3\Delta_2\rangle = -[2(X-2)]^{1/2}\,[B - \gamma/2 - A_D - 2D(X-1)].$$

$$\langle ^3\Delta_2|^3\Delta_3\rangle = -[2(X-6)]^{1/2}\,[B - \gamma/2 + A_D - 2D(X-5)].$$

$$\langle ^3\Delta_1|^3\Delta_3\rangle = -2[(X-2)(X-6)]^{1/2}D.\quad (3.21)$$

where $X = J(J+1)$. In eq (3.21), T is the energy of the vibrational level considered, B and D are rotational constants; A,ε and γ are the spin–orbit, spin–spin, and spin–rotation interaction parameters respectively. Explicit matrix elements in case (a) basis for $^3\Pi$, $^3\Delta$ and $^3\Phi$ states, calculated using "classical phenomenological Hamiltonian" have been presented by Shenyavskaya et al [96] in their paper on the analysis of the

$d^3\Phi \rightarrow a^3\Delta$ and $b^3\Pi \rightarrow a^3\Delta$ transitions in the FT emission spectrum of ScF.

An example of analysis of transitions involving electronic states of high multiplicity is the FT emission spectroscopy of the $B^6\Pi$–$X^6\Pi$ band system in the 1.2 μm region by Wallin et al [97] who have given the Hamiltonian and matrix elements for the $^6\Pi$ and $^6\sum$ states in Hund's case (a) basis.

3.1.2 Dunham Expression for Diatomic Energy Levels

In the reported high resolution FT work on diatomic molecules, several authors have treated the data using Dunham expression and instead of the customary spectroscopic constants for molecules, a set of Dunham coefficients are presented.

Dunham [98] calculated the energy levels of a vibrating rotator by using the WKB theory, which can be written in the form

$$E(v,J) = \sum_{i,j} Y_{ij} (v+1/2)^i [J(J+1)]^j \quad (3.22)$$

The Y_{ij} are coefficients which depend on molecular constants. The above equation for energy can be expanded into

$$E(v,J) = Y_{00}+Y_{10}(v+1/2)+Y_{20}(v+1/2)^2+Y_{30}(v+1/2)^3+Y_{40}(v+1/2)^4+\ldots$$

$$+[Y_{01} + Y_{11} (v+1/2) + Y_{21} (v+1/2)^2+ Y_{31} (v+1/2)^3] J(J+1)$$

$$+[Y_{02}+Y_{12}(v+1/2) + Y_{22} (v+1/2)^2 +\ldots] J^2 (J+1)^2$$

$$+[Y_{03} + \ldots\ldots] J^3 (J+1)^3 +\ldots\ldots \quad (3.23)$$

The full expressions for the various Dunham coefficients Y_{ij} in terms of spectroscopic constants are available [99]. If B_e/ω_e is small (for most molecules, the ratio is ~ 10^{-6} and for light molecules it increases to ~ 10^{-3}), then the Y_{ij} are approximated with spectroscopic constants as

$$\begin{array}{lll} Y_{10} \approx \omega_e & Y_{01} \approx B_e & Y_{21} \approx \gamma_e \\ Y_{20} \approx -\omega_e x_e & Y_{02} \approx -D_e & Y_{11} \approx -\alpha_e \\ Y_{30} \approx \omega_e y_e & Y_{03} \approx H_e & Y_{12} \approx -\beta_e \\ Y_{40} \approx \omega_e z_e & & \end{array} \quad (3.24)$$

The term Y_{00} is related to other spectroscopic constants and provides an addition to the zero–point energy of a diatomic anharmonic oscillator [ref 87, p. 109].

To obtain satisfactory fit to the high resolution data, Dunham expressions of the orders with i = 6, 4 and 2 are needed for the G_v, B_v and D_v parameters (i.e. Y_{i0}, Y_{i1} and Y_{i2}). Focsa et al [23] have fitted 3299 fluorescence lines of Br_2 $B^3\Pi_{0u+} \rightarrow X^1\sum_g^+$ system to Dunham expression (i = 0 to 6, j = 0 to 2). For the $X^2\sum_g^+$ ground state of BeD, Focsa et al [78] have reported 28 Dunham coefficients (Y_{ij} with i = 0 to 7, j = 0 to 4).

The isotopic dependence of the Y_{ij} coefficients has been discussed by Watson [100] and is given by

$$Y_{ij} \propto \mu^{-(i+2j)/2} \tag{3.25}$$

Where μ is the reduced mass of the isotopic atoms. A set of mass–independent coefficients can be defined by

$$U_{ij} = \mu^{(i+2j)/2}\, Y_{ij} \tag{3.26}$$

The spectroscopic data from different isotopic forms of a molecule can be combined to give a single equation [for e.g. ref 68, 100, 101 (p208)].

$$E(v,J) = \sum_{i,j} \mu^{-(i+2j)/2}\, U_{ij}\,(v+1/2)^i [J(J+1)]^j \tag{3.27}$$

Small corrections to this formula involving Born–Oppenheimer breakdown constants have been made and eq (3.27) becomes [100, 102]

$$E(v,J) = \sum_{i,j} \mu^{-(i+2j)/2}\, U_{ij}\,[1 + (m_e/M_A)\Delta_{ij}^A$$

$$+ (m_e/M_B)\Delta_{ij}^B](v+1/2)^i [J(J+1)]^j \tag{3.28}$$

where the two atomic centres are labeled as A and B; m_e, M and μ are electron, atomic and reduced masses, and Δ_{ij}^A and Δ_{ij}^B are the Born–Oppenheimer breakdown constants for atoms A and B. In the papers on the high resolution FT spectroscopy of InH and InD [102], and LiH and LiD [103,104] the mass–invariant Dunham constants U_{ij} have been reported.

3.1.3 Intensity of Ro–Vibrational Lines

The intensity of individual ro–vibrational lines in an absorption spectrum is given by [105]

$$S_{v'',J''}^{v',J'} = \frac{8\pi^3}{3hc} \cdot \nu_{v'',J''}^{v',J'} \left[\frac{N_{v'',J''}}{g_{J''}} - \frac{N_{v',J'}}{g_{J'}} \right] L_i \,|R_{v''}^{v'}(J)|^2 \tag{3.29}$$

where $\nu^{v',J'}_{v'',J''}$ is the frequency of the line, L_i is the Honl–London factor [ref. 87,p 208], $N_{v'',J''}$ is the ro–vibrational population of lower state given by

$$\frac{N_{v'',J''}}{g_{J''}} = \frac{N_{v''}}{Q_R} \exp[-hcB_v J(J+1)/kT] \tag{3.30}$$

with Q_R being the lower state rotational partition function, and

$$|R^{v'}_{v''}(J)|^2 = |R^{v'}_{v''}|^2 * F \tag{3.31}$$

Where $|R^{v'}_{v''}|^2$ is the value of rotaionless transition moment and F is the Herman–Wallis factor [106] (essentially a vibration–rotation interaction factor which influences line intensities in vibration–rotation bands). The integrated intensity S being proportional to pressure, the quantity S^0 (in cm^{-1}. sec^{-1}. atm^{-1}) is introduced such that

$$S = S^0 (p/p_0). \tag{3.32}$$

For a fundamental vibrational band (vibrational transition 1←0), the population of upper state being negligible, we have

$$S^0(m) = \frac{8\pi^3}{3hc} \cdot \nu_0(m)|m||R_{1\leftarrow 0}|^2 F(m).\frac{N_{v''}}{Q_R} \cdot \exp[-hcB_v m(m+1)/kT] \tag{3.33}$$

where m = J+1 for R–branch and m= –J for P–branch, and Herman–Wallis factor is given by

$$F(m) = 1+Cm+Dm^2 \tag{3.34}$$

Similar expressions for intensity in absorption for fundamental band of HCl have been given by Chackerian et al [107].

Imajo et al [76], while analyzing the FT emission spectrum of TiCl in the 420 nm region, have studied the intensity of the $^4\Gamma_{5/2}$–$X^4\Phi_{3/2}$ electronic transition of the molecule. The emission intensities of individual R and P branch lines are expressed by

$$I_{em} \propto \nu^4 L_i \exp[-hcB'J'(J'+1)/kT] \tag{3.35}$$

Civis et al [37] have analyzed the FT emission spectrum of NeH^+ in the ground electronic state $^1\Sigma^+$ corresponding to vibrational transitions 1–0, 2–1 and 3–2, between 1800 and 3000 cm^{-1}. They have given the expression for the integrated intensity I(ν) of a ro–vibrational transition observed in emission as

$$I(\nu) = A\nu^4 \exp(-E'/kT)|\langle\psi_{v',J'}|M(\xi)|\psi_{v'',J''}\rangle|^2 \tag{3.36}$$

Where the constant A depends on the number density of the upper vibrational state and M(ξ) is the dipole moment function. ξ is dimensionless and is $(r-r_e)/r_e$, where r_e is the equilibrium internuclear distance. The quantity

$$|\langle\psi_{v',J'}\,|\,M(\xi)\,|\,\psi_{v'',J''}\rangle|^2=\langle\psi_{v',J'=0}\,|\,M(\xi)\,|\,\psi_{v'',J''=0}\rangle^2\,|\,m\,|\,F_{v'v''}(m) \tag{3.37}$$

where |m| stands for Honl–London factor (When $\Lambda''=0$), and $F_{v'v''}$ (m) represents the Herman–Wallis factor given by eq (3.34), with m=J+1 for R and m = – J for P branches.

3.2 LINEAR POLYATOMIC MOLEUCLES

In this sub–section, the theoretical expressions for Hamiltonian, ro–vibrational energy levels, anharmonic interactions and *l*–type resonances, and intensity of ro–vibrational lines of linear polyatomic molecules with suitable examples are presented.

3.2.1 Hamiltonian for Linear Polyatomic Molecules

Several theoretical formulations of the Hamiltonian for rotational and vibrational energies have been reported. The Hamiltonian matrix has T = G(v,*l*) + F(v,*l*,T) as its diagonal elements. The off–diagonal elements represent various anharmonic interactions, anharmonic resonances, *l*–type interactions and *l*–type resonances (discussed later). Yamada et al [108] have given the effective Hamiltonian for a linear molecule as

$$\tilde{\mathbf{H}}=\sum_{m,n}\tilde{\mathbf{H}}_{mn} \tag{3.38}$$

where m and n indicate the power of the vibrational and rotational operators respectively. In a given vibrational state v, the Hamiltonian has been expressed as

$$\tilde{\mathbf{H}}=\hat{h}_d+\hat{h}_0+\hat{h}_2+\hat{h}_4 \tag{3.39}$$

where $\hat{h}_d$ represents the terms which are diagonal for the substates in a given vibrational state, and $\hat{h}_0,\hat{h}_2$ and $\hat{h}_4$ represent the terms which couple the substates with the selection rules Δk = 0, ±2 and ±4 respectively (k represents $\sum l_i$ where l_i are the vibrational angular momenta. Δk = 0 represents vibrational *l*–type doubling and interaction, while Δk = ± 2 represents rotational *l*–type doubling and interaction). Yamada et al [108] have given the expressions for each term of the r.h.s. of eq(3.39).

The effective Hamiltonian of Yamada et al [108] has been modified by Niedenhoff and Yamada [109] who have given the effective Hamiltonian and matrix elements for the vibration–rotation energy levels of a linear molecule in which two bending modes are simultaneously excited by any quanta.

Watson [110] has expressed the basic molecular Hamiltonian as expansion in the powers of vibrational and rotational operators

$$\mathbf{H} = \mathbf{H}_{20} + \mathbf{H}_{30} + \mathbf{H}_{40} + \ldots\ldots\ldots \text{(vibrational terms)}$$

$$+\mathbf{H}_{21} + \mathbf{H}_{31} + \mathbf{H}_{41} + \ldots\ldots\ldots \text{(Coriolis terms)}$$

$$+\mathbf{H}_{02} + \mathbf{H}_{12} + \mathbf{H}_{22} + \ldots\ldots\ldots \text{(rotational terms)} \tag{3.40}$$

where m and n in $\mathbf{H}_{mn}$ are the degrees in the vibrational and rotational operators. $\mathbf{H}_{20}$ is taken as the unperturbed Hamiltonian.

Bosch et al [62] have given the Hamiltonian operator for the linear triatomic molecule $CuCl_2$ in the X $^2\Pi_g$ ground state $(v_1, v_2^{\,l}, v_3)$ level as

$$\mathbf{H} = \mathbf{H}_{vib} + \mathbf{H}_{SO} + \mathbf{H}_{RT} + \mathbf{H}_{rot} + \mathbf{H}_{\Lambda d} + \mathbf{H}_{ld} + \mathbf{H}_{sr} \tag{3.41}$$

The seven contributions to the Hamiltonian are the vibrational energy (referred to zero point level), the spin–orbit interaction, Renner–Teller, rotational, lambda–doubling, and spin–rotation energies respectively. Actual expressions for each of the components have been given by Bosch et al [62]

3.2.2 Ro–Vibrational Energy

The unperturbed term values are written as

$$T(v,l,J) = G(v,l) + F(v,l,J) \tag{3.42}$$

The vibrational term values are given by (ref 111, p. 210, 371)

$$G(v,l) = \sum_i \omega_i (v_i + d_i/2) + \sum_{i \le j} x_{ij} (v_i + d_i/2)(v_j + d_j/2) + \sum_{t \le t'} g_{tt'}\, l_t l_{t'} \tag{3.43}$$

Where ω_i are the fundamental vibration wavenumbers, x_{ij} are anharmonic constants, d_i is the degree of degeneracy of vibration, and $g_{tt'}$ are additional anharmonic constants. The index t applies to degenerate modes with vibrational angular momentum l_t. When energy levels are referred to the lowest level, the vibrational term values are written as

$$G_0(v,l) = \sum_i \omega_i v_i + \sum_{i \le j} x_{ij} v_i v_j + \sum_{t \le t'} g^0_{tt'}\, l_t l_{t'} \tag{3.44}$$

In high resolution FTIR spectroscopy, additional terms are introduced in the expression for vibrational term values. For e.g., in the case of DCN, Mollmann et al [16] have used the expression

$$G(v,l) = \sum \omega_i (v_i + d_i/2) + \sum\sum x_{ij} (v_i + d_i/2) (v_j + d_j/2) + g_{22} l^2$$

$$+ \sum\sum\sum y_{ijk}(v_i+d_i/2)(v_j+d_j/2)(v_k+d_k/2) + \sum y_{ill} (v_i+d_i/2) l^2$$

$$+ \sum\sum\sum\sum z_{ijkh} (v_i+d_i/2)(v_j+d_j/2)(v_k+d_k/2)(v_h+d_h/2)$$

$$+ \sum\sum z_{ijll} (v_i+d_i/2)(v_j+d_j/2) l^2 + z_{llll} l^4 + z_{22222} (v_2+1)^5$$

$$+ z_{222ll} (v_2+1)^3 l^2 + z_{2llll} (v_2+1) l^4 \qquad (3.45)$$

where $G(v,l) = G_0(v,l) + G(0,0)$, with $G(0,0)$ representing the zero point vibrational energy. In eq (3.45), the sums run over 1,2 and 3 for the three normal vibrational modes of DCN with the condition that $h \geq k \geq j \geq i$.

The degeneracy of the normal modes is given by $d_1=d_3=1$ for the stretching modes and $d_2=2$ for the bending mode.

The rotational term values are given by [14]

$$F(v,l,J) = B_v [J(J+1)-l^2] - D_v[J(J+1)-l^2]^2 + H_v[J(J+1)-l^2]^3 \qquad (3.46)$$

In the case of HCN, the rotational constants are given by [14]

$$B_v = B_e - \sum \alpha_i (v_i+d_i/2) + \sum\sum \gamma_{ij} (v_i+d_i/2)(v_j+d_j/2) + \gamma_{ll} l^2$$

$$+ \sum\sum\sum \gamma_{ijk} (v_i+d_i/2)(v_j+d_j/2)(v_k+d_k/2)$$

$$+ \sum \gamma_{ill} (v_i+d_i/2) l^2 + \gamma_{2222} (v_2+1)^4 + \gamma_{2223} (v_2+1)^3 (v_3+1/2)$$

$$+ \gamma_{11ll} (v_1+1/2)^2 l^2 + \gamma_{22ll} (v_2+1)^2 l^2 + \gamma_{13ll} (v_1+1/2)(v_3+1/2) l^2$$

$$+ \gamma_{23ll} (v_2+1)(v_3+1/2) l^2 + \gamma_{llll} l^4 \qquad 3.47)$$

The centrifugal distortion coefficients for HCN are given by [14]

$$D_v = D_e + \sum \beta_i (v_i+d_i/2) + \sum\sum \beta_{ij} (v_i+d_i/2)(v_j+d_j/2) + \beta_{ll} l^2$$

$$+ \sum \beta_{ill} (v_i+d_i/2) l^2 \qquad (3.48)$$

and

$$H_v = H_e + \sum \varepsilon_i (v_i+d_i/2) + \varepsilon_{22} (v_2+1)^2 + \varepsilon_{ll} l^2 \qquad (3.49)$$

For Π vibrational states, the rotational term values for a triatomic molecule like BrCN are given by [52]

$$F(J) = (B_v \pm 0.5\, q_v)\, J(J+1) - (D_v \pm 0.5\, q_v^J)\, [J(J+1)]^2$$

$$+ H_v[J(J+1)]^3 \quad (3.50)$$

where q is the *l*–doubling constant such that

$$q = B(\Pi_f) - B(\Pi_e) \quad (3.51)$$

and

$$q^J = D(\Pi_f) - D(\Pi_e) \quad (3.52)$$

In the case of four–atomic molecules like HCCH, HCCX (with X = F, Br, I etc…) and their deuterated isotopic molecules, there are two bending vibrational modes ν_4 and ν_5, and each are doubly degenerate, giving rise to vibrational angular momenta l_4 and l_5.

The rotational term values are then given by [112]

$$F(v,J) = B_v\, [J(J+1) - k^2] - D_v[J(J+1) - k^2]^2 + H_v[J(J+1) - k^2]^3 \quad (3.53)$$

Where, the quantum number k, defining the component of J about the linear axis is given by

$$k = l_4 + l_5 \quad (3.54)$$

(note: some authors use *l* in place of k).

Holland et al [113] have analyzed the high resolution FTIR spectrum of the ν_4 band system of the linear molecule OCCCCCO (also called 1,2,3,4–penta tetraen–1,5–dione) and to fit the ro–vibrational data, used the energy expression

$$E_v = G_v + B_v\, [J(J+1)] - D_v[J(J+1)]^2 + H_v\, [J(J+1)]^3 + L_v\, [J(J+1)]^4 + M_v\, [J(J+1)]^5 \quad (3.55)$$

More refined treatment of a large number of experimental data (observed wavenumbers of ro–vibrational lines) requires more number of molecular parameters to be introduced in the Hamiltonian matrix, as mentioned in the case of HCN [14] and DCN [16]. The unperturbed ro–vibrational energies, obtained as diagonal elements, when both degenerate modes ν_4 and ν_5 of HCCH are excited, have been given by Herman et al [114]. The diagonal and off–diagonal elements of the Hamiltonian matrix formulated by Herman et al [114] have been widely used by several workers in the field of high resolution FT spectroscopy (for e.g., for $^{12}C_2H_2$ by Kabbadj et al [46], for $^{13}C_2H_2$ by Di Lonardo et al [47], and for HCCI and DCCI by Sarkkinen [115]).

For a five–atomic linear molecule like isocyanoacetylene (HCCNC), which has 7 vibrational modes (4 non–degenerate stretching vibrations ν_1 to ν_4 and 3 degenerate bending vibrations ν_5 to ν_7), the diagonal ro–vibrational energies E_{vr} containing terms involving E_v^0, B_v^0, D_v^0 and H_v^0 have been given by Vigouroux et al [116], and the expansion of each of the terms incorporating a number of vibrational, rotational and vibration–rotation interaction parameters have also been given.

3.2.3 Centre Energy, Band Centre and Band Origin

Centre energy or centre term value of a vibrational level is given by [112]

$$G_c = G(v)-(B_v k^2+D_v k^4)+........ \tag{3.56}$$

where k is given in eq (3.54).

The band centre wavenumber is defined as

$$\tilde{\nu}_c = G'_c - G''_c \tag{3.57}$$

and the band origin wavenumber is given by

$$\tilde{\nu}_0 = G'(v) - G''(v) = \tilde{\nu}_c + B'_v k'^2 - B''_v k''^2 + ... \tag{3.58}$$

3.2.4 Anharmonic Interactions and Resonances

As mentioned in section 1 (Introduction), molecular energy levels are perturbed due to processes like IVR and IVRET. Off–diagonal elements in the Hamiltonian matrix reepresent various anharmonic interactions and resonances, some of which will be discussed here. The mechanisms of IVR are vibrational resonances of Fermi type and of the Darling–Dennison type. Anharmonic interactions involving not necessarily Fermi or DD processes, but two sets of combination levels of same species, also occur, as in HCCCN.

3.2.4.1 Fermi Resonance

Fermi resonance is a perturbation involving a fundamental and an overtone or combination level of the same symmetry. In the case of a triatomic molecule like BrCN, Burger et al [52] have studied the strong Fermi resonance between ν_1 and $2\nu_2$ states. The Fermi resonance linking the $(02^00)_e$ and $(10^00)_e$ levels are given by

$$W_F = W_{12} + W_{12}^J J(J+1) \tag{3.59}$$

where W_{12} is the Fermi parameter and W_{12}^J gives the J(J+1) dependence of W_{12}.

Fermi resonance involving $v_1 = v_2=1$ and $v_2=3$ levels of CS_2 has been studied by Ahonen et al [117] who have described the interaction with the matrix element

$$\langle v_1, v_2^{\ell_2} \| (v_1 - 1), (v_2 + 2)^{\ell_2} \rangle = 1/2[W + W^J J(J+1)]$$

$$x\ \{v_1[(v_2 + 2)^2 - l_2^2]\}^{1/2} \qquad (3.60)$$

The $\nu_1(\sum_g^+)/\nu_2 + 2\nu_5(\sum_g^+)$ Fermi resorance in $^{13}C_2H_2$ has been studied by Di Lonardo et al [118] who have given the matrix elements of the interaction as

$$\langle v_1, v_2, v_5^{\ell_5}, J, k|H|(v_1-1), (v_2+1), (v_5+2)^{\ell_5}, J, k\rangle$$

$$= \{W_0 + W^J[J(J+1)-k^2]\}\ [v_1(v_2+1)\ (v_5+2)^2 - l_5^2]^{1/2} \qquad (3.61)$$

where W_0 is a coupling term whose value was found to be 3.134351 (35) cm^{-1} and the rotational dependence of the coupling term is W^J whose value was found to be 6.819 (24) x 10^{-5} cm^{-1} [118].

The Fermi resonance between ν_3 and $\nu_2+\nu_4+\nu_5$ levels of acetylene is represented by the matrix elements [85, 119]

$$\langle v_1, v_2, v_3, v_4^{\ell_4}\ v_5^{\ell_5},\ J\,|\,\widetilde{H}_{40}\,|\,v_1, (v_2+1), (v_3-1), (v_4+1)^{\ell_4 \pm 1}, (v_5+1)^{\ell_5 \mp 1}, J\rangle$$

$$= \frac{1}{8}\ K_{2345}\ [(v_2+1)\ v_3(v_4 \pm l_4+2)\ (v_5 \mp l_5+2)]^{1/2} \qquad (3.62)$$

where K_{2345} is called the "effective anharmonic constant" or the "Fermi resonance coupling constant", which is a complex function of potential constants [120]. The Fermi coupling term is $W_0 = \frac{\sqrt{2}}{4} K_{2345}$ for acetylene [85, 119].

Anharmonic resonances have been observed and analyzed in the FT spectra of linear molecules like HCCNC [116],HCCCN [121] and DCCCN [122]. Vigouroux et al [116] have given explicit expressions for the first order anharmonic resonance between ν_4 & $2\nu_6$, and between ν_4 and $\nu_5+\nu_7$ states; second order anharmonic resonance between ν_5 and $3\nu_7$ states, and third order anharmonic resonance between ν_4 and $4\nu_7$ states of HCCNC. Winther et al [121] have given the energy matrix for the anharmonic interaction of $\nu_1 + 2\nu_7^0$ with $\nu_4+3\nu_5+\nu_6+2\nu_7$ ($\sum^e$ sublevels) of HCCCN. Coveliers et al [122] have given expressions for the first order anharmonic interaction between ν_4 and $2\nu_6$, and for the third order anharmonic interaction between ν_4 and $4\nu_7$ states of DCCCN.

3.2.4.2 Darling–Dennison Resonance

The DD resonance involves the interaction of two overtone vibrations of same symmetry, while no interaction occurs between the two fundamental vibrations due to their different symmetries. In the case of linear molecules like C_2H_2 and C_2D_2, there are several DD resonances observed. Herman et al [123] have analyzed the $2\nu_1/2\nu_3$ DD coupling in $^{12}C_2D_2$

using an off–diagonal matrix element involving DD resonance constant K_{1133}. The x–K relations [124] predict that the value of K_{1133} must be close to the value of anharmonicity constant x_{13}. Herman et al [123] found that to reach a satisfactory result, both providing a low observed–calculated value and to obtain the x–K relations, an extra $\nu_1\nu_2/2\nu_3$ stretch–stretch DD resonance had to be included, involving constant K_{1233}. The values of K_{1133} and K_{1233} were determined as – 47.2 (14) cm^{-1} and – 54.4 (20) cm^{-1} for $^{12}C_2D_2$ [123].

Smith and Winn [125] have analyzed the DD resonances $2\nu_1/2\nu_3$ and $3\nu_3/(2\nu_1+\nu_3)$ in C_2H_2 by using an expression similar to that of Herman et al [123], and obtained the value of $|K_{1133}|$ to be 104.786 (6) cm^{-1}. Vaittinen et al [126] have analyzed the DD resonance between ν_1 and $\nu_2+2\nu_4$ states of HCCBr, have given the expression for matrix element, and have calculated the interaction parameter K_{1244} to be −14.96 cm^{-1}.

3.2.4.3 Coriolis Interactions

This type of interaction mixes vibrational and rotational modes of a molecule. As an example, Coriolis interaction in HCN couples levels that differ by one quantum of ν_3 and three quanta of ν_2 such as the levels $(01^11)_e$ and (04^00). The interaction couples levels that have same rotational J value but *l*–value differing by one. The *e* levels are coupled with *e* levels and *f* levels with *f* levels. Due to the rotational constants being different for such pairs of levels, there will be a crossing or near coincidence of levels at some value of J. Maki et al [127] have, in the FT absorption spectrum of HCN, observed Coriolis interaction involving $\Delta v_3=-1$, $\Delta v_2=3$ and $\Delta l=\pm1$ for many levels. The Coriolis coupling matrix element used by them is given by

$$\langle v_1, v_2, v_3, l\,|v_1,(v_2+3),(v_3-1), l\pm1\rangle = \mp W[J(J+1) - l(l+1)]^{1/2} \tag{3.63}$$

where W is the Coriolis interaction constant. Since the measured displacement of levels seemed to require two different values for the interaction constant, Maki et al [127] added a J–dependent term to the matrix element equation (3.63) such that

$$W = W_0 + W_J\, J(J+1) \tag{3.64}$$

Maki et al [14] analyzed a similar type of Coriolis interaction in the high temperature (1370K) FT emission spectrum of $H^{12}C^{14}N$, $H^{12}C^{15}N$ and $H^{13}C^{14}N$ by using expressions similar to eqs (3.63) and (3.64).

In the HCCBr spectrum, Coriolis interaction between high rotational levels of the close lying fundamentals ν_3 and ν_4 has been observed and analyzed by Vaittinen et al [128]. The matrix elements for the Coriolis interaction had the form [128]

$$\langle (v_3-1), (v_4+1), (l_4+1), J, k+1|H_{cor}/hc|v_3,v_4,l_4, J,k\rangle$$

$$= [B\xi_{34}^{(y)}\Omega_{34} + \xi_{34J}J(J+1)](v_3)^{1/2}\,(v_4+l_4+2)^{1/2}\,[J(J+1)-k(k+1)]^{1/2} \tag{3.65}$$

$$\langle (v_3+1), (v_4-1), (l_4+1), J, k+1\,|H_{cor}/hc|v_3,v_4,l_4,J,k\rangle$$

$$= -[B\xi_{34}^{(y)}\Omega_{34} + \xi_{34J}J(J+1)](v_3+1)^{1/2}\,(v_4-l_4)^{1/2}\,[J(J+1)-k(k+1)]^{1/2} \tag{3.66}$$

where k=l_4 (in the present case), and

$$\Omega_{34} = 1/2\ [(\tilde{\nu}_3/\tilde{\nu}_4)^{1/2} + (\tilde{\nu}_4/\tilde{\nu}_3)^{1/2}], \tag{3.67}$$

$\tilde{\nu}_3$ and $\tilde{\nu}_4$ are band origin wavenumbers for the ν_3 and ν_4 bands, $\xi_{34}^{(y)}$ is the Coriolis constant about the y axis (z is symmetry axis) and ξ_{34J} describes the J(J+1) dependence of $B\xi_{34}^{(y)}\Omega_{34}$. For the Coriolis coupled pair of vibrational states $(0001^10^0)\Pi_e$ and $(0010^00^0)\ \Sigma_e^+$ of HCC ^{79}Br, Vaittinen et al [128] have reported the values of $W_C = 2B\xi_{34}^{(y)}\Omega_{34} = 0.0128(70)\text{cm}^{-1}$ and $W_{CJ} = 2\xi_{34J} = -2.00(39)\text{x}10^{-7}\text{cm}^{-1}$

Coriolis resonance in DCCI has been investigated by Sarkkinen [115], using expressions identical to eqs (3.65) and (3.67) but without including the ξ_{34J} term, between ν_3 and ν_4, and between (00110) and (00020) levels. Explicit matrix elements for both types of Coriolis interactions have been given by Sarkkinen [115].

3.2.5 *l*–type Resonances

When doubly degenerate bending vibrations are excited in a linear molecule, vibrational angular momentum ***l*** is generated along the molecular axis. The coupling between levels that have the same quantum numbers but different *l* quantum numbers gives rise to *l*–type resonances. In the case of a linear triatomic molecule in which one quantum of ν_2 is excited, *l*–type resonance can be seen as a doubling of the states (the *e* and *f* levels of a Π vibrational state). When more than one quantum of ν_2 is excited, different values of *l* occur. Off–diagonal elements in the Hamiltonian matrix are used to represent *l*–type resonances. The effect of *l*–type resonance is of two categories, viz, off–diagonal elements involving *l* and *l*±2, and *l* and *l*±4.

In the case of a four atomic linear molecule, degenerate bending modes ν_4 and ν_5 exist. Ro–vibrational resonances occur which removes the degeneracy associated with *l* and results in a high density of interacting levels. Both vibrational and rotational *l*–type resonances occur in linear molecules with more than three atoms.

The theory of vibrational *l*–type doubling and resonance, and rotational *l*–type doubling and resonance in linear polyatomic molecules and the corresponding off–diagonal matrix elements have been given in a pioneering paper by Amat and Nielsen [129]. The formalism developed by Amat and Nielsen [129] has been used as a basis by several workers in the analysis of *l*–type resonances [for e.g., refs 108–110, 116,120]. The theory of vibrational and rotational *l*–type resonances has been given by Yamada et al [108], Niedenhoff and Yamada [109], Watson [110] and Pliva [120].

3.2.5.1 Triatomic Molcules

Maki and Mellau [15] have studied the *l*–type resonance effects in HNC and the off–diagonal matrix element for the (*l*, *l*±2) resonance is given as

$$\langle v_1, v_2, v_3, l, J|H/hc|v_1,v_2,v_3, l\pm 2, J\rangle$$

$$= \frac{1}{4} [q_v - q_{vJ} J(J+1) + q_{vJJ} J^2(J+1)^2] \{(v_2 \mp l)(v_2 \pm l+2)[J(J+1) - l(l\pm 1)]$$

$$\times [J(J+1) - (l\pm 1)(l\pm 2)]\}^{1/2} \quad (3.68)$$

where q_v, q_{vJ} and q_{vJJ} are l–type resonance constants, which have been determined for HNC upto v_2=5 [15]. For HCN, Maki et al [14, 127] have given matrix element similar to that in eq (3.68). The higher order term representing the (l,l–4) resonance in HCN has been given as [127]

$$W(l,l-4) = W(l-4,l) = (\rho/16) \{(v_2+l)(v_2-l+2)(v_2+l-2)(v_2-l+4)$$

$$\times [J(J+1) - l(l-1)] [J(J+1) - (l-1)(l-2)] [J(J+1) - (l-2)(l-3)]$$

$$\times [J(J+1) - (l-3)(l-4)]\}^{1/2} \quad (3.69)$$

The $(l,l\pm 4)$ coupling constant $\rho = -(0.1085\pm 0.0016) \times 10^{-7}$ cm^{-1} for $H^{12}C^{14}N$ [127]. Ahonen et al [117] have studied the l–resonance between $|l|$=1 and $|l|$=3 sublevels of v_2=3 in CS_2 using an expression similar to eq (3.68).

3.2.5.2 Four Atomic Molecules

As mentioned in section 3.2.2, in four–atomic molecules, the two degenerate vibrations v_4 and v_5 give rise to vibrational angular momenta l_4 and l_5 such that $k=l_4+l_5$. Vibrational l–type doubling and resonance, and rotational l–type doubling and resonance occur in these molecules.

(a) Vibrational l–type Resonances

In C_2H_2, for e.g., vibrational l–resonance effects of type $\Delta l_t = \pm 2$ and $\Delta l_{t'} = \mp 2$ occur with $r_{tt'}$ as the interaction parameter. The corresponding off–diagonal matrix elements are given by [114,120, 130]

$$\langle v_4^{\ell_4}, v_5^{\ell_5}, k \,\|\, v_4^{\ell_4\pm 2}, v_5^{\ell_5\mp 2}, k\rangle$$

$$= \frac{1}{4} r_{45}[(v_4 \mp l_4)(v_4 \pm l_4+2)(v_5 \pm l_5)(v_5 \mp l_5+2)]^{1/2} \quad (3.70)$$

Where r_{45} is the vibrational l–resonance parameter, the rotational and vibrational dependence of which is expressed as [47]

$$r_{45} = r_{45}^0 + r_{45}^J M + r_{45}^{JJ} M^2 \quad (3.71)$$

where r_{45}^0 represents the unperturbed value of r_{45} and M=J(J+1). Additional terms to the r.h.s. of eq (3.71) are also used [114,130].

(b) Rotational l–type resonances

As an example, there are three types of rotational l-resonance effects observed in C_2H_2, involving

a. $\Delta l_t = \pm 2, \Delta l_{t'} = 0$ with q_t as interaction parameter
b. $\Delta l_t = \pm 4, \Delta l_{t'} = 0$ with ρ_t as interaction parameter
c. $\Delta l_t = \pm 2, \Delta l_{t'} = \pm 2$ with $\rho_{tt'}$ as interaction parameter

with t,t′ = 4 or 5.

The corresponding off–diagonal matrix elements have been given by Herman et al [114]. q_t is the rotational l–resonance parameter whose rotational and vibrational dependence is given by [47, 114]

$$q_t = q_t^0 + q_{tt} v_t + q_{tt'} v_{t'} + q_t^J M + q_t^{JJ} M^2 \tag{3.72}$$

The parameter ρ_t is given by [47, 114]

$$\rho_t = \rho_t^0 + \rho_{tt} v_t + \rho_t^{t'} v_{t'} + \rho_t^J J(J+1) \tag{3.73}$$

Some corrections to the matrix elements in appendix B of Herman et al [114] have been given by Kabbadj et al [46]. ***l***–type resonances in other four atomic linear molecules have also been analyzed from FT spectra by several workers. Vibrational l-type resonances in HCCBr [126, 128] and HCCF [112], and rotational l-type resonances HCCBr [128], HCCF [112], HCCI [115, 131], DCCI [115, 132], NCCN [133] and C_2D_2 [123] are some examples of molecules in which analyses have been carried out.

3.2.5.3 Other Linear Molecules

(a) Vibrational l-type resonances

Vibrational l-type resonances in the FT spectra of linear molecules like HCCNC [116], HCCCN [121], DCCCN [122] and NCCCCN [134] have been analyzed. As an example, dicyanoacetylene (C_4N_2) has 9 modes of vibration (5 stretching vibrations ν_1 to ν_5 and 4 degenerate bending vibrations ν_6 to ν_9). Winther and Hegelund [134] have analyzed vibrational l-type resonances within each of the vibrational states (v_6, v_9) = (1,1), (1,2) and (1,3). The matrix elements used for the (ν_6+nν_9) vibrational system analysis are identical to those given in eq (3.70). A value for the l-type doubling parameter r_{69} =–0.0332 cm^{-1} for C_4N_2 was obtained by Winther and Hegelund [134].

(b) Rotational l-type resonances

Rotational l-type resonances in the FT spectra of molecules like HCCNC [116], DCCCN [122], SCCCS [135] and NCCCCN [134] have been analyzed. In the case of C_4N_2, the matrix elements for rotational l-type resonance are given by [134]

$$\langle v_t, l_t+1, J, k+1, M|H|v_t, l_t-1, J, k-1, M\rangle$$

$$= \frac{1}{4}[q_t + q_{tJ} J(J+1)][(v_t+1)^2 - l_t^2]^{1/2}$$

$$\times \{[J(J+1)-k(k+1)][J(J+1)-k(k-1)]\}^{1/2} \quad (3.74)$$

where l_t represents vibrational angular momentum corresponding to degenerate vibration t, $|J, k, M\rangle$ represents the rotational wavefunction, and q_t is the l-doubling constant whose J-dependent correction is given by q_{tJ} J(J+1).

3.2.6 Absolute Intensities of Bands in Linear Polyatomic Molecules

Intensity expressions for linear polyatomic molecules are similar to those for diatomic molecules given in eqs. (3.29)–(3.31), (3.34). However, there are some additional factors to be included. The individual line intensities are fitted to the formula [136, 137]

$$S_i = \frac{8\pi^3}{3hc} \cdot n \cdot \frac{273.15}{T} \cdot C \cdot |R_v|^2 \cdot \frac{\nu_i L_i}{g_v} \frac{\exp[-(E_v'' + E_r'')/kT]}{Q_v . Q_r} \times [1 - \exp(-hc\nu_i)/kT].F \quad (3.75)$$

where n is Loschmidt number = 2.68676 x 10^{19} cm^{-3}, T is measurement temperature, C is the isotopic abundance, R_v is the transition dipole for the vibration-rotation band, ν_i is the line frequency, L_i is the Honl-London factor, g_v is the degeneracy of levels involved, E_v'' and E_r'' are the lower state vibrational and rotational energies respectively, Q_v and Q_r are the lower state vibrational and rotational partition functions respectively, k is the Boltzmann constant, and F is the Herman–Wallis factor (originally introduced as correction factors for vibration-rotation interaction for diatomic molecules but subsequently extended to both linear and non-linear polyatomic molecules).

For perpendicular bands of a linear molecule, the Honl-London factors are [136]

$$L_i^P = (J-1-l\Delta l)\,(J-l\Delta l)/2J \quad (3.76)$$

$$L_i^Q = (J+1+l\Delta l)\,(J-l\Delta l)\,(2J+1)/2J(J+1) \quad (3.77)$$

$$L_i^R = (J+2+l\Delta l)(J+1+l\Delta l)/2(J+1) \quad (3.78)$$

for the P, Q and R branches.

The Herman-Wallis factors for the above branches are [138]

$$F^{RP} = [1 + A_1^{RP} m + A_2^{RP} m^2]^2 \quad (3.79)$$

and

$$F^{Q} = [1 + A_2^{Q}\ J(J+1)]^2 \tag{3.80}$$

where m=–J for P and m = J+1 for R-branches.

The band intensities S_v are given by [137]

$$S_v = \frac{8\pi^3}{3hc} \cdot n \cdot \frac{273.15}{T} \cdot C \cdot |R_v|^2 .\nu_0 \frac{.\exp(-E''_v / kT)}{Q_v} \cdot [1 - \exp(-h\nu_0 / kT)] \tag{3.81}$$

where ν_0 is the band origin.

The expressions and explanations for different quantities involved in line strength, transition dipole matrix elements and vibrational band intensity etc. have been given by Grecu et al [139] in their paper on the determination of absolute ro-vibrational intensities of ν_5 absorption band system of NCCN.

Chapter 4

TYPES OF INFORMATION OBTAINED FROM FOURIER TRANSFORM SPECTROSCOPY

In this section, the types of information obtained from high resolution FT spectroscopy of linear molecules are presented.

4.1 DIATOMIC MOLECULES

The largest contribution in the field of high resolution FT spectroscopy of linear molecules is from diatomic molecules, mostly in emission and few in absorption; and hence the types and examples of the information obtained from FT experiments on diatomic molecules will be covered in more detail in this review.

4.1.1 Rotational Analysis of Bands

An example of a high resolution FT spectrum of a diatomic molecule is shown in fig. 4.1 which is of the (0,0) band of $B^1\sum^+ \rightarrow X^1\sum^+$ transition of CuCl reported by Parekunnel et al [79]. The figure shows the P and R-branches, the band head and the band origin. Parekunnel et al [79] measured the ro-vibronic line positions for several bands with transitions from the electronic states $a\,^3\sum_1$, $b\,^3\Pi_1$, $b^3\Pi_0$, $A^1\Pi$ and $B\,^1\sum^+$ to the ground state, in the 18000–25000 cm^{-1} region, recorded at a resolution of 0.02 cm^{-1}.

Faye et al [140] have analyzed the bands of the $B^3\Pi_g$–$W^3\Delta_g$ system of $^{14}N_2$, recorded in the 1250–2250 cm^{-1} region with an unapodized resolution of 0.0043 cm^{-1}. The B and W states have potential energy curves very close to each other, due to which, overlapping transitions B→W and W→B are observed. The schematic potential curves of B and W states of N_2 are shown in fig 4.2. The origin of the energy is the v=0 level of the $B^3\Pi_g$ state. The transitions observed by Faye et al [140] are indicated in the figure, which shows that the flow of energy passes through one electronic state to another, illustrating an intrasystem cascading. A part of the 1→0 band of B→W transition of N_2 is shown in fig 4.3 in which three Q-branch lines and one P-branch line are shown.

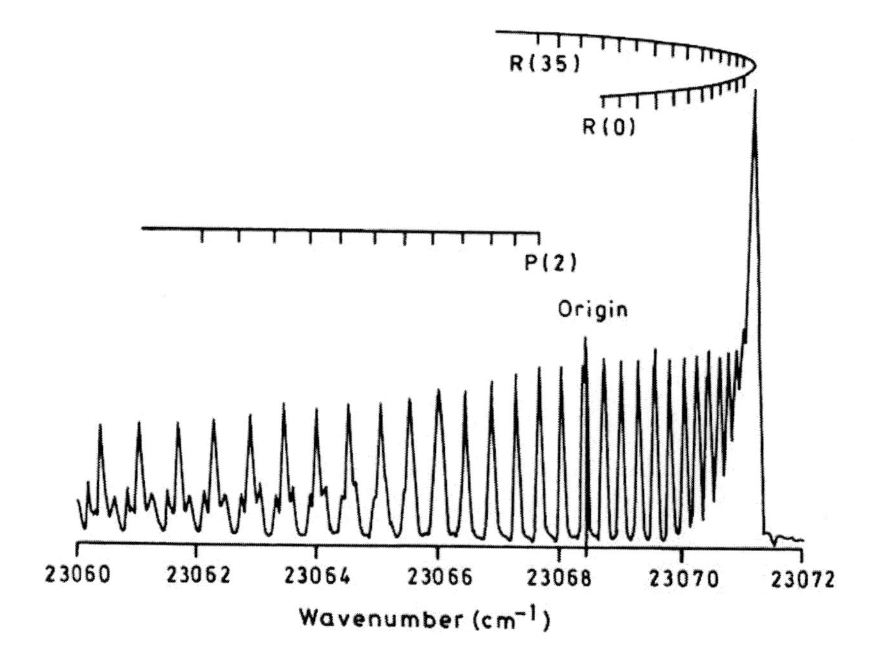

Figure 4.1 The (0,0) band of the $B\,^1\Sigma^+ - X\,^1\Sigma^+$ transition in the emission spectrum of CuCl. The figure shows the band origion and the band head in the R–branch (Reproduced from Parekunnel et al [79] with permission from Elsevier Science & Academic Press, Orlando, Florida, USA).

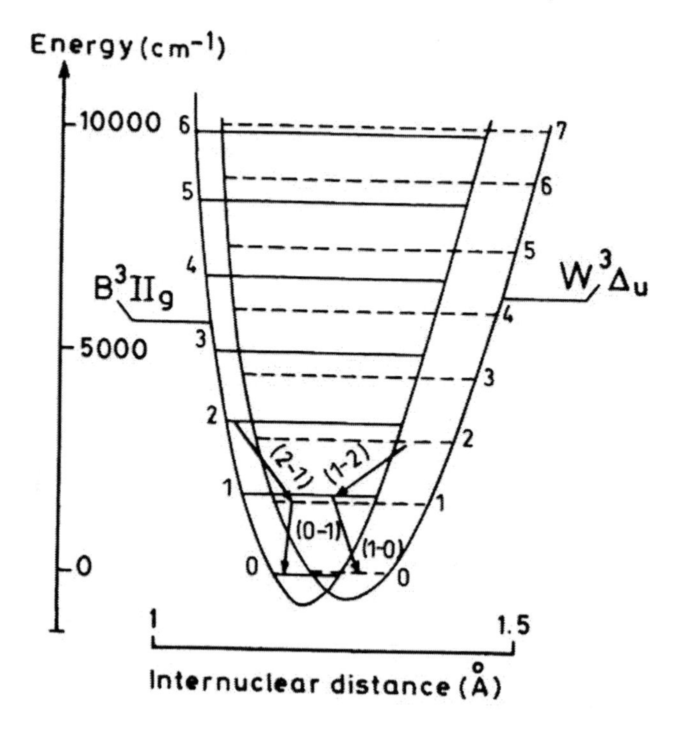

Figure 4.2 Schematic diagram of potential energy curves showing the intrasystem cascading between the $B^3\Pi_g$ and $W^3\Delta_u$ states of $^{14}N_2$. The origin of energy is taken in the $B^3\Pi_g$ v= 0 level. Vibrational transitions investigated are also shown (Reproduced from Faye et al [140] with permission from Elsevier Science & Academic Press, Orlando, Florida, USA).

The method of optical-optical double resonance (OODR) coupled with FTS has been used by Ross et al [141] to study the $2^1\Sigma_u^+$ "double minimum" state of Li_2. With a CW dye laser, the strong $A^1\Sigma_u^+ \leftarrow X^1\Sigma_g^+$ (v=0–4) transition was excited, and another dye laser was used to reach the Rydberg $5d^1\Pi_g$ state from the $A^1\Sigma_u^+$ state. Infrared emission corresponding to $5d\ ^1\Pi_g \rightarrow C^1\Pi_u$ and $5d^1\Pi_g \rightarrow 2^1\Sigma_u^+$ transitions were analyzed by Ross et al [141], who have reported the turning points for the inner well of the $2^1\Sigma_u^+$ state.

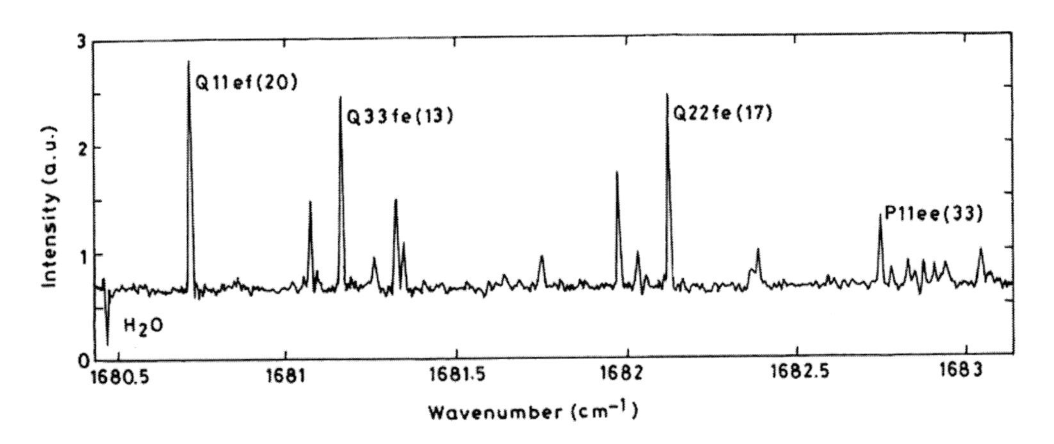

Figure 4.3 A part of the N_2 emission spectrum in the (1,0) band `region of the $B^3\Pi_g$–$W^3\Delta_u$ systemSome Q–branch lines and one P–branch line are shown in the figure (Reproduced from Faye et al [140] with permission from Elsevier Science & Academic Press, Orlando, Florida, USA).

4.1.2 Fine Structure Transitions

The $^2\Pi_i$ ground state of tellurium monohalides show splitting identified as $X_2\ ^2\Pi_{1/2}$ and $X_1\ ^2\Pi_{3/2}$. The near infrared emission spectrum of $X_2 \rightarrow X_1$ fine structure transitions of TeF and TeCl were analyzed by Ziebarth et al [142] who had reported the spectroscopic constants for the v=0 levels of both molecules, and found that in spite of the large spin-orbit splitting, the rotational levels of the two components of $X\ ^2\Pi_i$ states can be described equally well in both Hund's cases (a) and (c) formalisms and that the case (a) approach is preferable.

Fink et al [143] have analyzed the X_21–X_10^+ fine structure transition of BiF at an apodized resolution of 0.005 cm^{-1}. The X_21 and X_10^+ are the fine structure components of $X\ ^3\Sigma^-$ ground state of BiF. Two magnetic hyperfine parameters were determined for BiF involving its ground state. They are the Fermi contact term b_F = –0.0249(10) cm^{-1} and the dipole-dipole coupling constant c = –0.1912 (10) cm^{-1}.

Shestakov et al [67], in an extensive paper, have reported the spectrum of fourteen new electronic transitions of BiO radical. The fine structure transition $X_2{}^2\Pi_{3/2} \rightarrow X_1{}^2\Pi_{1/2}$ was found to yield the strongest emission in the near infrared region (fig 1m. of ref. 67).

The $X_2{}^2\Pi_{3/2}\rightarrow X_1{}^2\Pi_{1/2}$ fine structure transitions of PbF and PbCl have been analyzed by Ziebarth et al [19] who have reported the accurate spectroscopic constants for the ground states of PbF and PbCl, and also the hyperfine structure constants d = –0.242 (1) cm^{-1} and d_D = –5 (8) x $10^{-7}cm^{-1}$ for ^{207}PbF.

4.1.3 Magnetic Dipole Transitions

The intensity of magnetic dipole transitions are weaker by a factor of 10^{-5} than the intensity of electric dipole transitions. Since FT spectra could be recorded with high sensitivity and high S/N ratio, such weak magnetic dipole transitions could be detected.

Fink et al [144], while analyzing the $b^1\Sigma^+–X^3\Sigma^-$ spectrum of SeS, observed the $b0^+–X_10^+$ and $b0^+–X_21$ subsystems of the molecule, where $\Omega=0^+$ and 1 substates of X are the case (c) components of $^3\Sigma^-$ state. In the (0,0) band of $^{80}Se^{32}S$, four electric dipole branches and three magnetic dipole branches were observed.

Setzer et al [38] have analyzed the high resolution FT emission spectrum of SO molecule, due to the $b^1\Sigma^+–X^3\Sigma^-$ and $a^1\Delta\rightarrow X^3\Sigma^-$ transitions. SO is isoelectronic with O_2 and has $X^3\Sigma^-$ ground state. The strong (0,0) band of the $b^1\Sigma^+–X^3\Sigma^-$ transition could be measured with high S/N ratio, which showed five electric dipole branches ^SR, ^QR,^QQ,^QP and ^OP, and four magnetic dipole branches ^RQ, ^RR, ^PP and ^PQ. A section of the spectrum of (0,0) band of the $b^1\Sigma^+–X^3\Sigma^-$ system of SO is shown in fig. 4.4, where the rotational assignments for the ^RR and ^RQ magnetic dipole branches are indicated, along with the electric dipole branches ^SR, ^QP and ^QR. From the relative line intensities and the radiative lifetime of the $b^1\Sigma^+$ state, Setzer et al [38] determined the electronic transition moment to be $\mu_0 = \pm$ 0.0042 ea_0 and $\mu_1 = \mp$ 0.0047 ea_0. The magnetic transition moment M was determined to be $|M| = 0.16\ \mu_B$. The square of the transition moment is connected with μ_0, μ_1 and M by the relation [38]

$$|R^e_{nm}|^2 = \mu_0^2+2\mu_1^2+2M^2 \tag{4.1}$$

The electric and magnetic transition moments μ_0, μ_1 and M appear in Watson's rotational line strength formulas [table II, ref. 145].

A purely magnetic dipole fine structure transition, the $X_21_g\rightarrow X_1 0_g^+$ transition, between the fine structure components of the ground state of Pb_2 was observed by Setzer et al [146]. The Δv=0, +1 and –1 sequences in the above fine structure transition were observed in the region from 5199.53 cm^{-1} to 5460.29 cm^{-1} by Setzer et al [146].

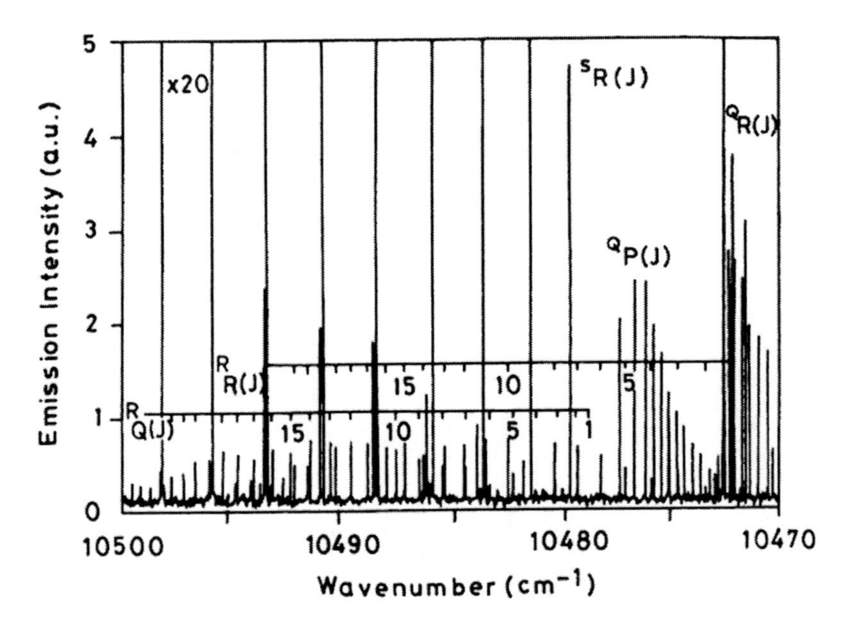

Figure 4.4 A section of the expanded spectrum of the (0,0) band of the $b^1\sum^+ \rightarrow X^3\sum^-$ system of SO. The rotational assignments for the ^{R}R and ^{R}Q magnetic dipole branches are indicated, alongwith the electric dipole branches ^{S}R, ^{Q}P and ^{Q}R (Reproduced from Setzer et al [38] with permission from Elsevier Science & Academic Press, Orlando, Florida, USA).

4.1.4 Potential Energy Curves

Potential energy curves for ground and excited states of many diatomic molecules have been constructed and reported by several workers. The experimentally obtained values of vibrational energy and rotational constants (or Dunham coefficients) from FTS experiments are used to generate turning points, potential energy parameters and potential energy curves.

Fig. 4.5 shows the Rydberg–Klein–Rees (RKR) potential curve between v=0 and v=13 for $^{39}K^{85}Rb$ $(2)^3\sum^+$ state obtained by Amiot [63]. The calculated values of Rousseau et al [147] are also shown in the figure. The bottom of the potential curve shown in fig. 4.5 corresponds to a term value of 13507.1464 cm^{-1} which is the value of T_e of the $(2)^3\sum^+$ state. The $(2)^3\sum^+$ state of KRb dissociates into K(4s)+Rb (5p) atomic states, and the dissociation energy of the molecular state has been estimated to be 3447 cm^{-1} by Amiot [63].

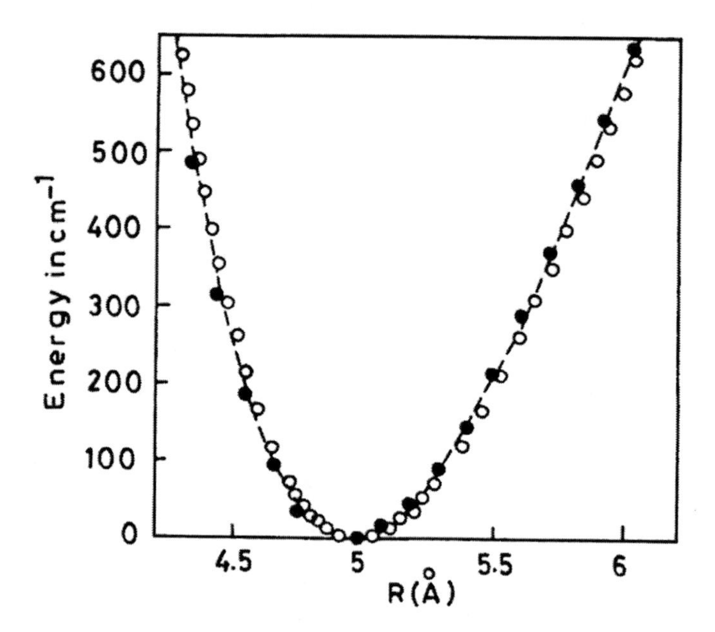

Figure 4.5 RKR potential energy curves of the $^{39}K^{85}Rb$ $(2)^3\Sigma^+$ electronic state (filled circles), derived from the molecular constants determined from the analysis of the $(3)^1\Pi \rightarrow (2)^3\Sigma^+$ system of KRb molecule. The energy origin is for the hypothetical level v=–1/2, J=0. Calculated values of the turning points from Rousseau et al [147] are represented by open (Reproduced from Amiot [63] with permission from Elsevier Science & Academic Press, Orlando, Florida, USA).

RKR turning points and RKR potential energy curves for electronic states of molecules reported from FTS experiments include the $X^3\Sigma^-$ state of PH [148], $(5)^1\Pi$, $(7)^1\Sigma^+$ $(4)^1\Pi$, $(2)^1\Pi$ and $(3)^1\Sigma^+$ states of RbCs [149]; $X\ ^1\Sigma_g^+$, $A^1\Pi_u$, $B^1\Delta_g$ and $B'\ ^1\Sigma_g^+$ states of C_2 [150, 151]; $B^3\Pi_g$, $A\,^3\Sigma_u^+$ and $B'\ ^3\Sigma_u^-$ states of $^{15}N_2$ [152] ; $a^1\Pi_g$ and $w\ ^1\Delta_u$ states of $^{14}N_2$ [153]; $A^2\Pi_i$ and $X\,^2\Sigma^+$ states of CP [154]; $a^3\Pi$ and $b\,^3\Sigma^-$ states of BH [155], and $X\ ^1\Sigma_g^+$ state (v=0–29) of Br_2 [23]. Amiot et al [156] have reported the RKR potential curves for the $X\ ^2\Sigma^+$, $A'\ ^2\Delta$, $A^2\Pi$, $B\,^2\Sigma^+$, $C^2\Pi$, $E\,^2\Sigma^+$ and $F\,^2\Sigma^+$ states of BaCl. The RKR turning points of $X0_g^+$ state (upto v=108) of $^{127}I_2$ [157]; $X0_g^+$ states (upto v=108) of $^{129}I_2$ and (upto v=109) of $^{127}I\ ^{129}I$ [158]; $^1\Sigma^+$ ground state (upto v=119) of ^{85}RbCs [159]; $A^1\Sigma^+$ state of NaLi [160]; $X\ ^1\Sigma^+$ and $B^1\Pi$ (*e*–parity component) of ScF [161] and $X\ ^1\Sigma^+$ state (upto v=66) of NaK [162] are also the examples of reported work.

The "Inverted perturbation approach" (IPA) potential curves (turning points) constructed from molecular constants obtained from FTS experiments have been reported for the $X\ ^1\Sigma^+$ state of CO [163], $1^1\Pi_g$ state (upto v = 68) of Rb_2 [164] and also $2\,^1\Sigma_g^+$ state (upto v = 25) of Rb_2 (165]. Quite recently, Park et al [24] have calculated and drawn the potential energy curves for the electronic states of Rb_2 by "Multireference configuration interaction using the averaged relativistic effective small-core potential and core-polarization potential". The potential energy curves calculated for the $1^1\Pi_g$ and $2\,^1\Sigma_g^+$ states of Rb_2 have been compared by Park et al [24] with the experimentally (LIF+FTS) obtained potential curves of the same states by Amiot [164], and Amiot and Verges [165] respectively; and found that the agreement between theory and experiment was perfect.

Lee et al [66] have reported the parameters of the "expanded Morse oscillator" (EMO) function for the ground state $X\,^1\Sigma^+$ of $^{74}Ge^{16}O$, according to whom, the combined isotopomer "Direct potential fit analysis" is more compact and more physically significant alternative to the Dunham type analysis and that the resulting potential energy funciton is quantum mechanically accurate. Potential parameters and potential energy curves obtained from "Direct potential fit" to the multi-isotopomer data sets for $X\,^1\Sigma^+$ and $A\,^1\Sigma^+$ states of ^{63}CuH, ^{107}AgH and ^{197}AuH have been reported by Seto et al [166].

Hedderich et al [167] have reported the energy parameters obtained by fitting the FT emission spectral data to the Born-Oppenheimer potential (which is a modified Morse potential function, given in eq 6 of ref. 166) and potential curve for the $X\,^1\Sigma^+$ state of AlCl. Born-Oppenheimer potential parameters and potential energy curves for the $X\,^1\Sigma^+$ state of AlH [168] and $X\,^1\Sigma^+$ state of GaH [169] have also been reported.

4.1.5 Pressure Broadening, Shifts, and Asymmetries in Bands

Pressure induced broadening and lineshifts are important in atmospheric studies. FT spectrometers provide extremely high wavenumber accuracy over a wide range and hence reliable pressure shift and pressure broadening data are available. Self broadening, broadening induced by other atoms/molecules, self shift, shifts induced by other atoms/molecules, and asymmetry are some of the pressure induced effects reported. Many authors report band strengths and intensities of lines alongwith the results on the pressure induced effects.

When the recorded spectral lines do not overlap, they are represented by the normalized profile [29]

$$I(\omega) = \frac{1}{\pi} \{[\gamma + Y(\omega - \omega_0 - \delta)] / [(\omega - \omega_0 - \delta)^2 + \gamma^2]\} \qquad (4.2)$$

Where γ is the linewidth, δ is the lineshift and Y is the asymmetry parameter. These parameters vary linearly with the number density of the perturbing atom/molecule. The pressure shift at a single pressure leads to asymmetric lineshape. Grigoriev et al [29] have reported the expressions to evaluate the broadening and shifting cross sections.

NO plays an important role in the chemistry of the troposphere and stratosphere. N_2-broadened linewidths of NO fundamental band [170], N_2 induced broadening and shift coefficients for the $^2\Pi_{1/2}$ and $^2\Pi_{3/2}$ subbands of the fundamental band [171], temperature dependence of the N_2-broadening coefficients $\gamma(T)$ for several transitions of the fundamental band [172] and O_2-broadening coefficients (polynomial representations) for the fundamental band [33] are some examples of the reported work. Recently, Pope and Wolf [32] have reported the rotational level dependent broadening coefficients for the $^2\Pi_{1/2}$ and $^2\Pi_{3/2}$ subbands of the NO (1←0) band using five noble gases (He, Ne, Ar, Kr, Xe) as broadening species. Fig. 4.6 shows the example of a linear least squares fit to the Lorentzian HWHM versus argon pressure for the R (6.5) lines of NO reported by Pope and Wolf [32]. Subtraction of the self broadening contribution values determined by Ballard et al [table IIa, ref. 170] caused of the intercept to pass through origin. The slopes of the lines provided the broadening coefficients. The behaviour of the broadening coefficients for P and R branches as a function of rotational level showed strong dependence on rotational quantum number, which indicated angular momentum dependence of the inelastic collision mechanism.

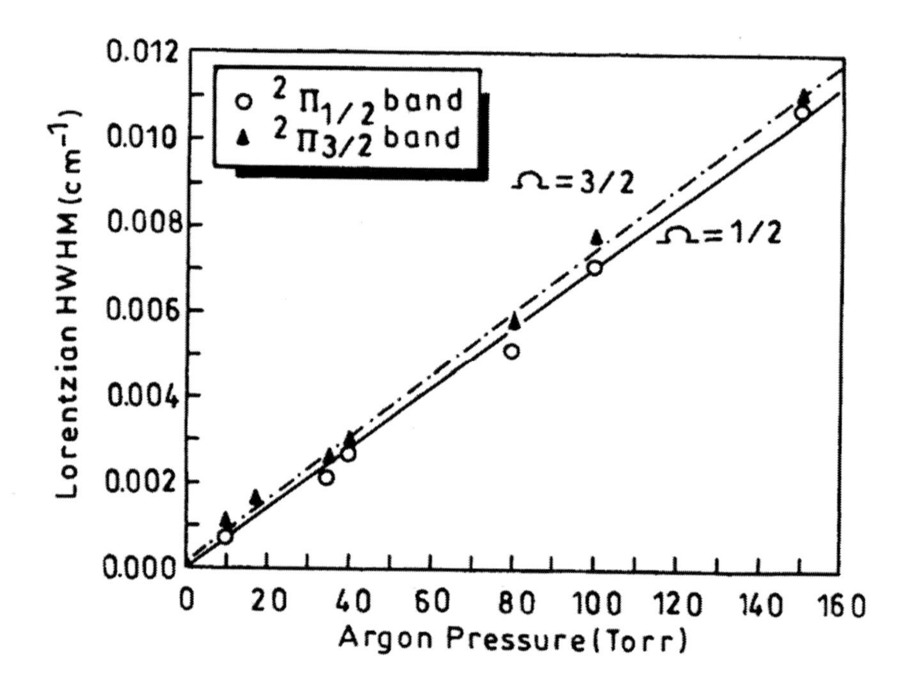

Figure 4.6 Linear least squares fit to the Lorentzian HWHM as a function of argon bath pressure. The data belongs to the R(6.5) lines of the $^2\Pi_{1/2}$ and $^2\Pi_{3/2}$ substates of NO fundamental band, broadened by Ar. The slope of these lines provide the pressure broadening coefficients (Reproduced from Pope and Wolf [32] with permission from Elsevier Science & Academic Press, Orlando, Florida, USA).

CO is one of the most important atmospheric trace gases and a study of its concentration gives idea about the variable mixing ratios in the troposphere, stratosphere and mesosphere. Bouanich et al [173] have determined the pressure induced shift coefficients in the CO fundamental band perturbed by He, Ne, Ar, Xe, O_2 and N_2, and found that both lineshift and broadening coefficients, depended on $|m|$. Voigt et al [147] have reported the N_2 broadening of the $^{13}C^{16}O$ (2←0) band and the Lorentzian halfwidths of the lines were found to increase linearly with total pressure. Chackerian et al [28] have reported the self broadening and self shift coefficients for the (3←0) band of CO, and the individual rovibrational intensities were estimated by assuming a Voigt–Galatry lineshape H(X,Y,Z). Fig 4.7 shows the observed wavenumber of the m = –14 line versus total gas pressure in the 3←0 band of CO [28]. The slope of the line gave self shift coefficient (–672(43) x 10^{-5} cm^{-1}/atm) and the intercept of the plot yielded the vacuum wavenumber (6393.176740(14) cm^{-1}) for the m = –14 line.

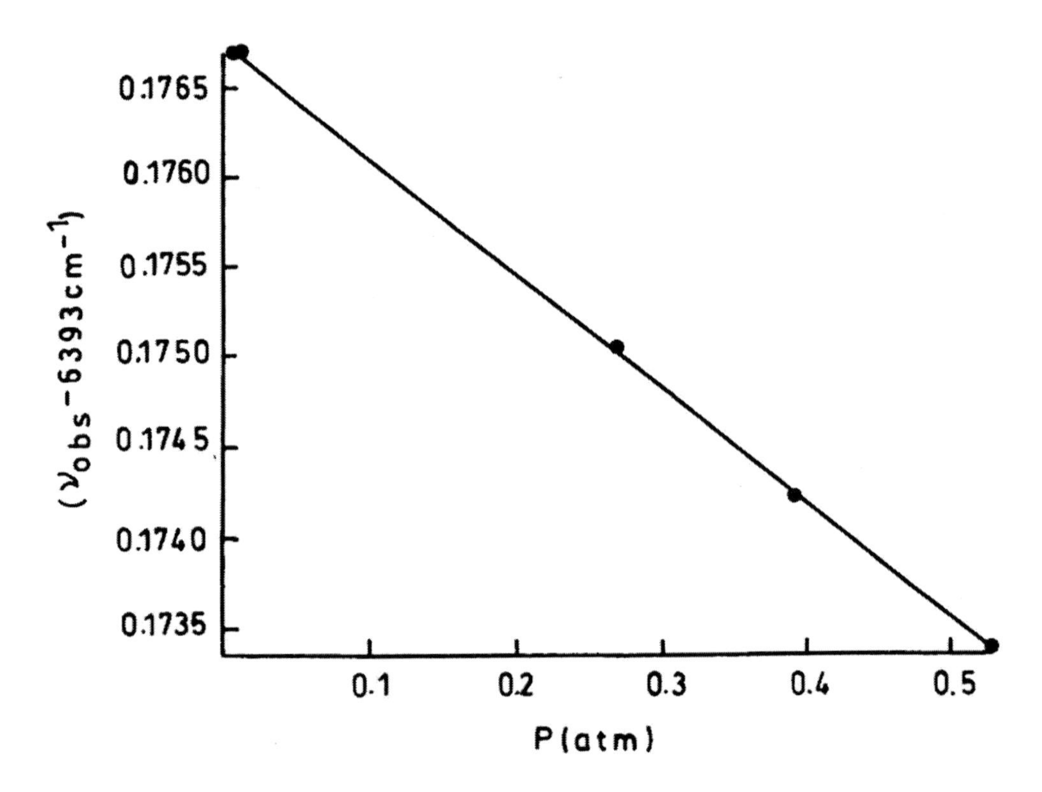

Figure 4.7. The observed wavenumber of the m= –14 line of $X^1\sum^+$ $v = 3 \leftarrow v = 0$ band of $^{12}C^{16}O$ vs pressure. The pressure shifts were measured from the spectra recorded at a resolution of 0.0109 cm^{-1} (Reproduced from Chackerian Jr et al [28] with permission from Elsevier Science & Academic Press, Orlando, Florida, USA).

The infrared absorption spectrum and collision broadening in HF is also of importance in the study of the terrestrial stratosphere. Nitrogen broadened halfwidths of HF lines in the fundamental band by Thompson et al [175] and measurement of widths and shifts in the (0–0), (0–1) and (0–2) bands of HF in a bath of argon by Grigoriev et al [29] to yeild the Ar-HF broadening and shifting cross sections, are some examples of reported work. Agreement between the experimental cross sections by Grigoriev et al [29] and the theoretical values of close coupling cross sections reported by Green and Hutson [176] has been excellent.

The asymmetries of the profiles of the argon-broadened HF lines in the (0–0), (0–1) and (0–2) bands have been measured by Boissoles et al [31] who found that the asymmetries in the (0–0) band are weak, observable at very high densities, and are due to line mixing effects; while the asymmetries observed in the (0–1) and (0–2) bands at moderate densities can be attributed to non-impact effects through vibrational dephasing process.

4.1.6 Integrated Cross Sections and Band Oscillator Strengths

Accurate parameters like integrated cross sections and band oscillator strengths obtained from FT absorption intensity measurements of molecules like O_2 and NO are very important in the study of solar absorption in the terrestrial middle atmosphere. In the wavelength region above 205nm, oxygen molecules absorb solar radiation, dissociate into $O(^3P)$ atoms and subsequently produce O_2 and O_3. Some examples of FT spectroscopy applied to determine the integrated cross sections and band oscillator strengths are presented here.

The absorption cross sections are obtained by [177] using Beer's law

$$N\sigma(\lambda) = \ln [I_0(\lambda)/I(\lambda)] \tag{4.3}$$

where I_0 is the background intensity and N is the gas column density. The integrated cross sections of the bands are obtained by adding up all line cross sections belonging to each band. The band oscillator strengths are given by [177]

$$f(v',v'') = \frac{mc^2}{\pi e^2 \tilde{N}(v'')} \int \sigma(v)dv \tag{4.4}$$

where $\tilde{N}(v'')$ is the Boltzmann population of the absorbing vibrational level. The integration of cross section is performed over all of the rotational lines belonging to the (v′,v″) band.

The absorption spectrum of O_2 in the 33000–41700 cm^{-1} region consists of discrete Herzberg I,II and III band systems due to the $A^3\Sigma_u^+ - X^3\Sigma_g^-$, $c^1\Sigma_u^- - X^3\Sigma_g^-$ *and* $A'^3\Delta_u - X^3\Sigma_g^-$ transitions respectively, the Herzberg continuum, and the broad collision induced Wulf bands. Yoshino et al [177, 178] have reported the integrated cross sections and band oscillator strengths of the Herzberg II (v′=6–16) [177] and Herzberg III (v′=4–11) [178] band systems. Fally et al [41] have reported the Herzberg continuum cross sections and the collision induced absorption cross sections of the diffuse Wulf bands, and have pointed out that the Wulf bands arise from O_2–O_2 collisions, and are due to an enhancement of the forbidden $A'^3\Delta_u \leftarrow X^3\Sigma_g^-$ transition caused by the relaxation of dipole–forbidden selection rule during collision. Using a 50 m base path length multiple reflection cell and paths of 201.84, 402.08 and 602.32m with O_2 pressures ranging from 20 to 750 Torr in their absorption experiments, Merienne et al [40] have reported the integrated cross sections and band oscillator strengths for the Herzberg I (v′=0–11), Herzber II (v′=2–19) and Herzberg III (v′=2–12) bands.

Fig 4.8 shows the integrated cross sections for the $^QQ_{11}$, $^QQ_{22}$, and $^QP_{32}$ lines of the $A^3\Sigma_u^+ - X^3\Sigma_g^-$ (5,0) band of O_2 reported by Merienne et al [40]. The experimental points and the fitted curves according to the model of Bellary and Balasubramanian [179] are also shown in the figure.

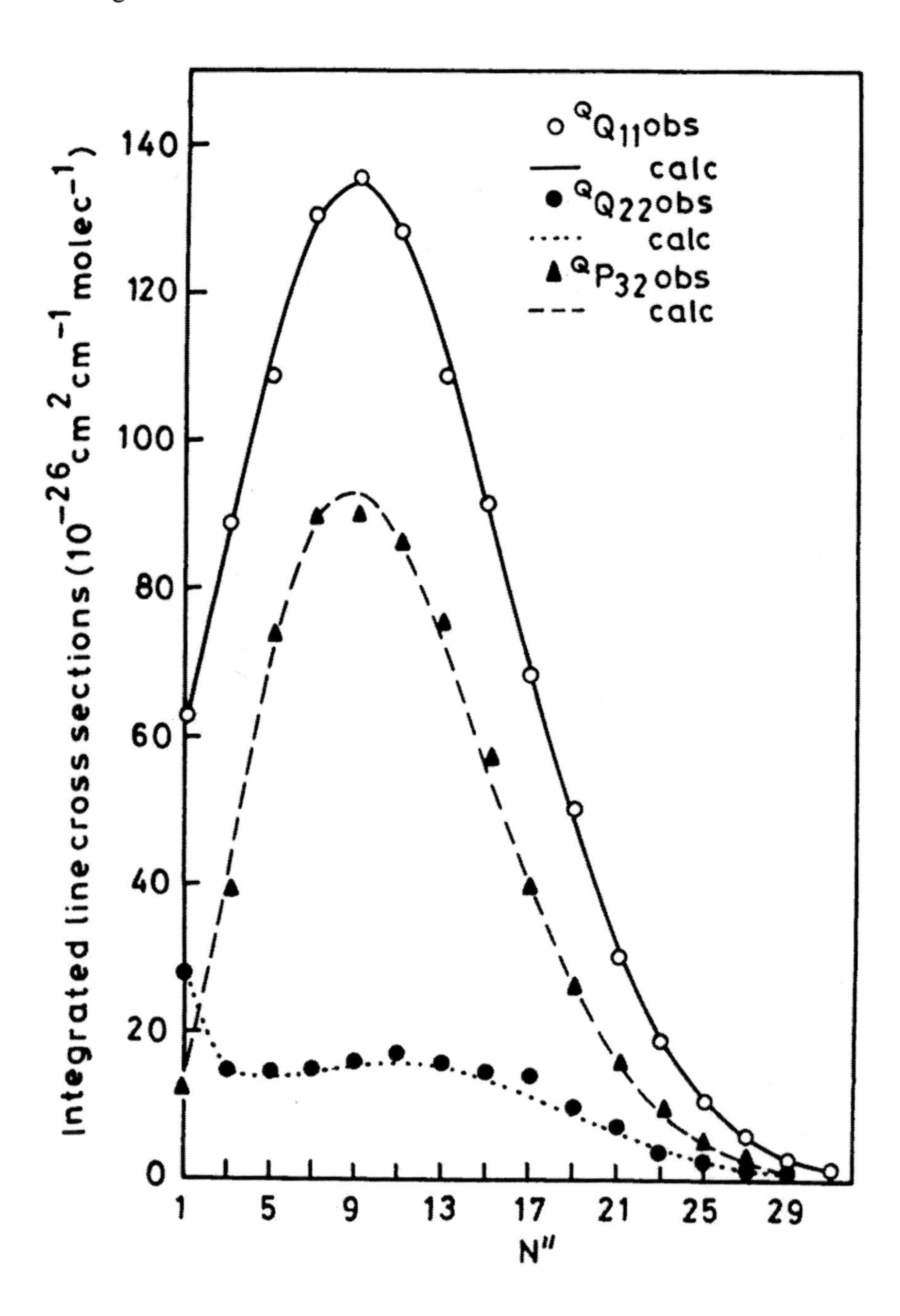

Figure 4.8. Integrated line cross sections for the $^QQ_{11}$, $^QQ_{22}$ and $^QP_{32}$ lines of the $A^3\sum_u^+ \leftarrow X^3\sum_g^-$ (5,0) band of O_2. The experimental values and fitted curves according to the model of Bellary and Balasubramanian [179] are shown (Reproduced from Merienne et al [40] with permission from Elsevier Science & Academic Press, Orlando, Florida, USA).

Using vacuum ultraviolet Fourier transform spectrometer and synchrotron radiation as a continuum source, the integrated cross sections of lines and band oscillator strengths for the β(9,0) [ref. 180], δ(1,0) [ref. 181], ε(1,0) [ref 182], and β(6,0) and γ(3,0) [ref. 183] bands of NO have been reported recently. The γ, β, δ and ε bands of NO correspond to

$$A\,^2\Sigma^+ - X\,^2\Pi_r,\; B\,^2\Pi_r - X\,^2\Pi_r,\; C\,^2\Pi_r - X\,^2\Pi_r \text{ and } D\,^2\Sigma^+ - X\,^2\Pi_r$$

transitions respectively. Among the above reported bands of NO, the δ(1,0) band has the highest value of band oscillator strength $f=(5.4\pm0.3) \times 10^{-3}$, while the β(6,0) band has the lowest value of $f = (0.48\pm0.05) \times 10^{-4}$.

4.1.7 Time Resolved FT Spectroscopy

Durry and Guelachvili [184] recorded the emission spectrum of N_2 $B\,^3\Pi_g \rightarrow A\,^3\Sigma_u^+$ transition from a pulsed microwave excited plasma and measured the temporal evolution of the intensity of the Δv = 0,1,2 and 3 sequences, and among other results, observed that the Δv=1 line intensity rises faster. TRFTS has been used by Lindner et al [185] to study the emission spectrum of vibrationally excited OH ($X^2\Pi$,v) formed in the reactions of ethyl, n–propyl and i–propyl radicals with $O(^3P)$ atoms. In the C_2H_5+O reactions, the total OH emission was found to increase with time and the v=3 state was most populated among the detected (v=1–5) OH levels. TRFTS is useful to study time varying phenomena.

4.1.8 Quadrupole Band Spectra

Nitrogen is a homonuclear diatomic molecule and hence electric dipole pure rotation and vibration–rotation spectra within the $^1\Sigma_g^+$ ground electronic state are forbidden. Rinsland et al [186] have recorded the high resolution stratospheric solar absorption spectra and have made accurate measurements of the positions of the O and S branch lines of the (1←0) vibration–rotation quadrupole band of $^{14}N_2$.

4.1.9 Detection of Valence States

By recording the emission spectrum of CS in the 9500–17500 cm^{-1} range, Choe et al [187] have studied the anomalous Λ–doubling observed in the $d^3\Delta_1$–$a^3\Pi_0$ (7,0) band, which allowed them to locate a previously unknown $^1\Sigma^-$ valence state (T = 40835.5 cm^{-1}, B = 0.597 cm^{-1}). The anomalous Λ–doubling was explained as due to a perturbation occurring in the $d^3\Delta_1$ (v=7) level. The origin of the $a^3\Pi_0$ (v=13) level is located just above (~12 cm^{-1}) the origin of the $d^3\Delta_1$ (v=7) level and these two levels may interact through the rotation–

electronic operator, and hence the $^3\Delta_1$ level may possess a significant fractional $^3\Pi_0$ character. If a level of the $^1\Sigma^-$ valence state is located below and near the preceding two levels, it can push the *f*–parity levels of the $^3\Delta_1$ state above those of *e*–parity. The $^3\Pi_0$ and $^1\Sigma^-$ levels are coupled via a spin–orbit interaction. Thus a triple interaction $^3\Delta_1{\sim}^3\Pi_0{\sim}^1\Sigma^-$ is considered to be responsible for the anomalous Λ–doubling in the $d^3\Delta_1$–$a^3\Pi_0$ (7,0) band [fig 6, ref. 187].

4.1.10 Location of Missing Electronic States

Due to the high sensitivity and large resolving power of the FTS technique, transitions from states which could not be detected by other experimental techniques, are observed. An example is the case of the $d^1\Sigma^+$ state of NbN. Langhoff and Bauschlicher Jr [188] had calculated the singlet and triplet states of niobium nitride below 20,000 cm^{-1} and predicted the ordering of the states in the singlet manifold to be $a^1\Delta$, $b\,^1\Sigma^+$, $c\,^1\Gamma$, $d\,^1\Sigma^+$ and $e^1\Pi$. All these states had been detected experimentally except the $d\,^1\Sigma^+$ state near 13000 cm^{-1}. Azuma et al [189] could not detect the $d\,^1\Sigma^+$ state in their laser excitation and emission experiments (and hence did not include the state in their figure 1, showing the electronic states and transitions of NbN). Recently, Ram and Bernath [73] have located the $d\,^1\Sigma^+$ state for the first time by observing the $d\,^1\Sigma^+$–$A\,^3\Sigma_0^-$ and $d\,^1\Sigma^+$–$b\,^1\Sigma^+$ transitions in the FT emission experiments. The $d\,^1\Sigma^+$ state was found to have T_0 = 13908.2989 (13) cm^{-1} by Ram and Bernath [73] who have also reported a portion of the $d\,^1\Sigma^+$–$b\,^1\Sigma^+$ (0,0) band of NbN near the P head [fig 2, ref. 73].

4.1.11 Observation of Deviation from Boltzmann Distributions

An example is provided in the FT emission experiments of Rehfuss et al [190] who produced supersonically cooled CN radicals in a corona discharge of acetonitrile seeded in an inert carrier gas (helium or argon). A non–Boltzmann and multi–temperature rotational distribution was found in the (0,0) band of the $B\,^2\Sigma^+ - X\,^2\Sigma^+$ transition. The vibrational population in the B–X transition was also found to deviate quite markedly from Boltzmann distribution for v′>6 [fig 6, ref. 190]. These results are expected to provide insight into the CN production mechanisms and its relaxation dynamics.

4.1.12 Molecular Constants and Parameters from High Resolution Spectra of Diatomic Molecules

In high resolution FT spectroscopy, it is customary to report a large number of molecular constants to fit the data very accurately, so that the standard deviation of the fit is very small.

Tables 4.1 and 4.2 show the accurate molecular constants reported by Gutterres et al [21] for the $X\ ^2\Sigma^+$, $B\ ^2\Sigma^+$ and $D\ ^2\Sigma^+$; and for the $A'^2\Delta_{3/2}$, $A^2\Pi$ and $C^2\Pi$ electronic states respectively for BaI. From tables 4.1 and 4.2 it can be seen that different sets of molecular parameters are required to describe the $^2\Sigma^+$ and $^2\Pi$ states. Gutterres et al [21] had derived the molecular constants from a simultaneous treatment of a data set of 12,684 transitions of BaI (with a standard deviation of 3.26 x 10^{-3} cm^{-1} of the fit) in which, the LIF + FTS results of Gutterres et al [21,69,94,191] and the selectively detected LIF (SDLIF) results of Leach et al [192] were used. The constants and parameters used to describe the $^2\Sigma^+$ states are introduced in equations (3.15)–(3.18), while those used to describe $^2\Pi$ states of BaI (table 4.2) are given equations (3.19)–(3.20) and in the table showing matrix elements for $^2\Pi$ states in section 3.

Table 4.1 Molecular constants in cm–1 for the $X\ ^2\Sigma^+$, $B\ ^2\Sigma^+$ and $D\ ^2\Sigma^+$ electronic states of BaI, determined in the analysis from a nonlinear least–squares fit of the global data set (numbers in parentheses represent two standard deviations in units of the last figure quoted) (Reproduced from Gutterres et al [21] with permission from Elsevier Science & Academic press, Orlando, Florida, USA).

Coefficient	$X\ ^2\Sigma^+$	$B\ ^2\Sigma^+$	$D\ ^2\Sigma^+$
T_e^a	[0]	10427.02271(58)	25775.11128(114)
B_e x 10^2	2.6804551(465)	2.6113433(466)	2.8088703(894)
D_e x 10^9	3.31357(571)	3.52044(578)	2.9373(216)
H_e x 10^{16}	−1.640(231)	−2.853(234)	−8.535(758)
α_B x 10^5	6.635018(987)	7.24250(113)	7.2232(210)
β_B x 10^8	3.4177(663)	5.0182(563)	−67.51(182)
γ_B x 10^{11}	8.21(172)	12.39(258)	781.7(511)
α_D x 10^{12}	1.5323(563)	2.1921(655)	82.56(400)
β_D x 10^{12}	–	–	−2.673(202)
ω_e	152.163175(272)	141.951547(292)	161.3901991(512)
$\omega_e x_e$	0.2726744(230)	0.2896530(276)	0.36421456(101)
$\omega_e y_e$ x 10^4	2.37212(590)	3.11148(776)	−17.0374409(340)
γ_e x 10^3	2.53620(222)	−56.40752(401)	2.123762652(14)
γ_J x 10^{10}	−3.749(208)	116.781(581)	−0.0557462(892)
γ_{JJ} x 10^{14}	–	–	3.693(389)
γ_v x 10^5	−1.1230(215)	8.2267(310)	3.536(208)
γ_{vv} x 10^6	–	–	−2.231(165)

[a]Origin of the energies at the level v=–1/2, N = 0 of the ground state.

Instead of the customary molecular parameters, several authors have reported the Dunham coefficients Y_{ij} (presented in equation (3.22)) derived from their high resolution experimental data. The equivalence of the various Y_{ij} to the spectroscopic constants have been given in equation (3.24). Table 4.3 shows the Dunham coefficients for the $X^1\Sigma^+$ state of RbCs reported by Fellows et al [61]. Fellows et al [61] had used 23,432 transitions taken from their LIF+FTS results and from those of Gustavsson et al [159], in their linear least squares fit (with root mean square error of 2.95 x 10^{-3} cm^{-1}) to obtain the Dunham coefficients (i = 1–13 and j=0–3), one of the largest set of coefficients reported so far from FTS work.

Table 4.2 Molecular constants in cm–1 for A'2Δ, A2Π, and C2Π electronic states of BaI determined in the analysis from nonlinear least–squares fit of the global data set (numbers in parentheses represent two standard deviations in units of the last figure quoted) (Reproduced from Gutterres et al [21] with permission from Elsevier Science & Academic press, Orlando, Florida, USA)

Coefficient	$A'\,^2\Delta_{3/2}$	$A\,^2\Pi$	$C\,^2\Pi$
T_e [a]	8369.0381(143)	9605.423748(652)	18188.50700(43)
B_e x 10^2	2.623438(122)	2.59470737(109)	2.6726955(493)
D_e x 10^9	5.0929(311)	3.46379(588)	3.05971(594)
H_e x 10^{16}	93.84(278)	–2.454(250)	–1.425(233)
α_B x 10^5	6.93524(231)	7.0235(355)	6.359995(986)
β_B x 10^8	4.4887(972)	3.4830(539)	2.7166(438)
α_D x 10^{12}	2.579(117)	2.561(137)	1.8343(593)
A_e x 10^{-2}	–	6.5665507(105)	7.56059904(667)
A_J x 10^6	–	–37.5060(163)	–4.0333 (187)
A_{JJ} x 10^{12}	–	–7.396(757)	4.6673(610)
A_v x 10^1	–	–5.24796(287)	1.09706(202)
A_{vv} x 10^3	–	2.3199(134)	1.5993(140)
A_{vJ} x 10^7	–	1.4055(767)	3.8049(154)
ω_e	142.285880(908)	141.747957(410)	157.795574(387)
$\omega_e x_e$	0.2741056(628)	0.2753768(455)	0.2747645(522)
$\omega_e y_e$ x 10^4	2.6083(189)	2.4094(147)	2.1759(227)
p_e x 10^2	–	–5.61110(160)	0.703091(554)
p_J x 10^8	–	1.3272(481)	–0.32291(493)
p_v x 10^5	–	7.226(158)	–4.3510(410)
q_e x 10^5	–	2.587(193)	–0.2022(450)
q_J x 10^{11}	–	–	1.374(284)
q_v x 10^6	–	2.682(706)	

[a]origin of the energies at the level v=–1/2, N = 0 of the ground state.

4.2 Triatomic Molecules

High resolution FT spectroscopy of a number of linear triatomic molecules have been reported in the literature during the past several years. A survey of the papers show that voluminous work has been done on CO_2, HCN and their isotopic variants. A number of papers on N_2O, OCS, OCSe, CS_2 and other molecules exist. In this section, the CO_2 molecule will be reviewed in more detail and HCN, N_2O, OCS, OCSe and CS_2 molecules in lesser detail. The experimental results for HCN and DCN will be quoted to highlight the types of molecular parameters obtained from high resolution work. Other linear triatomic molecules will be discussed in brief.

Table 4.3 Dunham–type coefficients for the $X\,^1\Sigma^+$ state of the RbCs (Reproduced from Fellows et al [61] with permission from Elsevier Science & Academic press, Orlando, Florida, USA).

i	J	Y_{ij}(cm^{-1})	%σ	i	j	Y_{ij}(cm^{-1})	%σ
1	0	0.5001373815 x 10^{2}	0.001	10	1	−0.1331930517 x 10^{-20}	8.700
2	0	−0.1095291172 x 10^{0}	0.045	11	1	0.2134560579 x 10^{-23}	9.313
3	0	−0.1210766696 x 10^{-3}	3.444	0	2	−0.7287292795 x 10^{-8}	0.125
4	0	−0.1617315317 x 10^{-5}	12.892	1	2	−0.4032678271 x 10^{-10}	1.857
5	0	0.2411457582 x 10^{-7}	27.033	2	2	0.3236914285 x 10^{-12}	29.498
6	0	−0.9318276740 x 10^{-9}	13.867	3	2	−0.1109820339 x 10^{-12}	9.203
7	0	0.1594306720 x 10^{-10}	9.777	4	2	0.7828106043 x 10^{-14}	7.617
8	0	−0.1321344712 x 10^{-12}	7.250	5	2	−0.2782018593 x 10^{-15}	6.776
10	0	0.8358859352 x 10^{-17}	4.564	6	2	0.5471723124 x 10^{-17}	6.184
11	0	−0.4899016229 x 10^{-19}	3.746	7	2	−0.6092324779 x 10^{-19}	5.658
13	0	0.5107958809 x 10^{-24}	2.540	8	2	0.3581008041 x 10^{-21}	5.172
0	1	0.1660058923 x 10^{-1}	0.001	9	2	−0.8663367528 x 10^{-24}	4.699
1	1	−0.3661023801 x 10^{-4}	0.087	0	3	0.2316265653 x 10^{-14}	4.154
2	1	−0.1983042834 x 10^{-6}	2.707	1	3	−0.3913428720 x 10^{-16}	14.852
3	1	0.4781087643 x 10^{-8}	10.879	3	3	0.5193124261 x 10^{-18}	9.810
4	1	−0.4234215672 x 10^{-9}	7.053	4	3	−0.4467261260 x 10^{-19}	8.096
5	1	0.1629450553 x 10^{-10}	6.574	5	3	0.1522626276 x 10^{-20}	7.263
6	1	−0.3911365624 x 10^{-12}	6.452	6	3	−0.2613924190 x 10^{-22}	6.550
7	1	0.5995397260 x 10^{-14}	6.694	7	3	0.2219999617 x 10^{-24}	5.903
8	1	−0.5938388969 x 10^{-16}	7.220	8	3	−0.7469989853 x 10^{-27}	5.260
9	1	0.3698809005 x 10^{-18}	7.942				

Note: %σ represents the error for each coefficient in percentage.

(A) CO_2 and its Isotopic Variants

For CO_2, the harmonic frequencies are related by $\omega_1 \approx 2\omega_2$ and $\omega_3 \approx 3\omega_2$. In the literature, the usual ($v_1v_2^lv_3$) as well as the HITRAN [193] notations for the vibrational levels of CO_2 are used. In the HITRAN notation, the energy levels are labeled as (v_1 v_2 l_2 v_3 r, $1 \leq r \leq v + 1$) with $v_2 = l_2$, based on the polyad structure. Each polyad is made up of vibrational basis states such that, polyad number $P = 2v_1 + v_2 + 3v_3$, and r is the ranking index of the Fermi interacting levels (starting from the highest). The polyads are labeled with integer P and the series of bands are labeled with the difference $\Delta P = P'–P$ where P′ and P are integers numbering the upper and lower polyads respectively. Both conventional and HITRAN notations for CO_2 will be used in this review.

(a) Absorption

Several bands of isotopically enriched carbon dioxide and of the unsymmetric species have been reported [194–196]. Using a modified Bomem DA 3.002 FT spectrometer with maximum optical path difference of 454 cm and an unapodized resolution of 0.0014 cm^{-1}, Tan et al [197] have reported several molecular parameters from analysis of the $10^00 \leftarrow 01^10$ and $02^20 \leftarrow 01^10$ bands of CO_2 with an absolute accuracy of ± 0.00007 cm^{-1} for the measured lines. Using the long path difference FT interferometer built by Valentin [48], Claveau et al [198] have analyzed the spectra of several isotopic species of carbon dioxide in the regions of the fundamental bands ν_1,ν_2 and ν_3; and spectroscopic constants have been reported for several hot bands. Using carbon dioxide enriched with oxygen–17, the line positions and molecular constants for 00011–10001, and 00011–10002 laser bands, near 10μm and 9μm have been reported by Claveau et al [199].

The strong Fermi coupled doublet (ν_1,$2\nu_2$) is forbidden in absorption by the rule of mutual exclusion, but has been observed in pressurized CO_2 as a "collision induced absorption" doublet by Barabov and Vigasin [34a]. The CIA spectrum in the vicinity of the lower and upper components of the Fermi dyad showed almost identical shape, consisting of pronounced Q-branches surrounded by P– and R–like branches [34a]. Quite recently, temperature variations in the CIA spectra of CO_2 in the region of the Fermi doublet (ν_1, $2\nu_2$) have been examined by Vigasin et al [34b]. Tashkun et al [200], in the global fit of the vibrational- rotational line positions of the $^{16}O^{12}C^{18}O$ and $^{16}O^{12}C^{17}O$ molecules, have used a number of FT experimental data and few laser diode data, to derive 73 parameters of a reduced effective Hamiltonian describing all vibrational rotational energy levels in the ground electronic state of the former molecule and 45 parameters of the latter molecule.

(b) Emission

A number of papers authored/co–authored by Bailly have been reported on the FT emission spectra of carbon dioxide. Vibrational transitions with $\Delta v_3 = 1$ of $^{12}C^{16}O_2$ involving $\sum$, Π, Δ and Φ groups were reported by Bailly et al [201] where the levels studied did not involve in Fermi resonance. The wavenumbers and spectroscopic constants for emission bands with $\Delta v_3 = 1$ of $^{12}C^{16}O_2$ involving the Fermi dyads $(10^0v_3, 02^0v_3)_{I,II}$ and $(11^1v_3, 03^1v_3)_{I,II}$ were reported by Bailly and Rossetti [202]. Other reported work include emission in the 4.5 μm region of $^{12}C^{16}O_2$ [203], $^{13}C^{16}O_2$ and $^{13}C^{18}O_2$ [204] $^{12}C^{16}O^{18}O$ [205]; observation of $\sum - \sum$ and Π–Π transitions in $^{13}C^{16}O^{18}O$ in a $^{13}CO_2$–N_2–O_2–He DC discharge [206], emission in the 15 μm region of $^{12}C^{16}O_2$ with $v_3 = 0$ [207], emission in the 15μm region involving $v_1' v_2'^{\ell'} v_3' \rightarrow v_1 v_2^l v_3$ (v_3 = 1,2,3) transitions [208], and the determination of rotational temperature T_R from the intensity of ro–vibrational lines in the CO_2 emission spectra from CH_4+O_2 low pressure flames [209] in which very low intensity lines with $2v_1+v_2=5$ were analyzed. Bailly et al [210] have studied the time evolution of the populations of excited vibrational levels of CO_2 and found that the 00^0v_3 levels (with v_3 upto 9) were observed for short times <100μs, whereas transitions involving combination levels $0v_2^l\ v_3$ (upto v_2 = 6) reach their maximum intensity for times > 1600 μs. The $(21^11)_3\ e \rightarrow (21^10)_3\ e$ transition had shown large observed–calculated value [fig 2, ref. 210]. Bailly et al [210] followed a path C_2 (from three possible paths constructed by using wavenumbers from the

HITRAN database) to compute the $(21^11)_3$ *e* ro–vibrational energy levels, which are perturbed by a Coriolis resonance with $(4000)_2$ ro–vibrational levels.

Bailly [211] analyzed the $^{12}C^{16}O_2$ emission in the 4.5μm region involving transitions $v_1v_2^l\ v_3 \rightarrow v_1v_2^l\ (v_3-1)$ with $2v_1+v_2 = 6$, excited in a DC discharge through CO_2+N_2 and observed 13 vibrational transitions occurring between highly excited vibrational states (very low populations involved) in the Fermi resonance polyads. A part of the emission spectrum of $^{12}C^{16}O_2$ in the range 2249.15–2249.80 cm^{-1} reported by Bailly [211] is shown in figure 4.9. Some P(J) lines in the $\Delta v_3 = -1$ transitions and the upper vibrational levels $(v_1v_2^lv_3)_n$ are shown in the figure, where n is the sequence number of the Fermi resonance group.

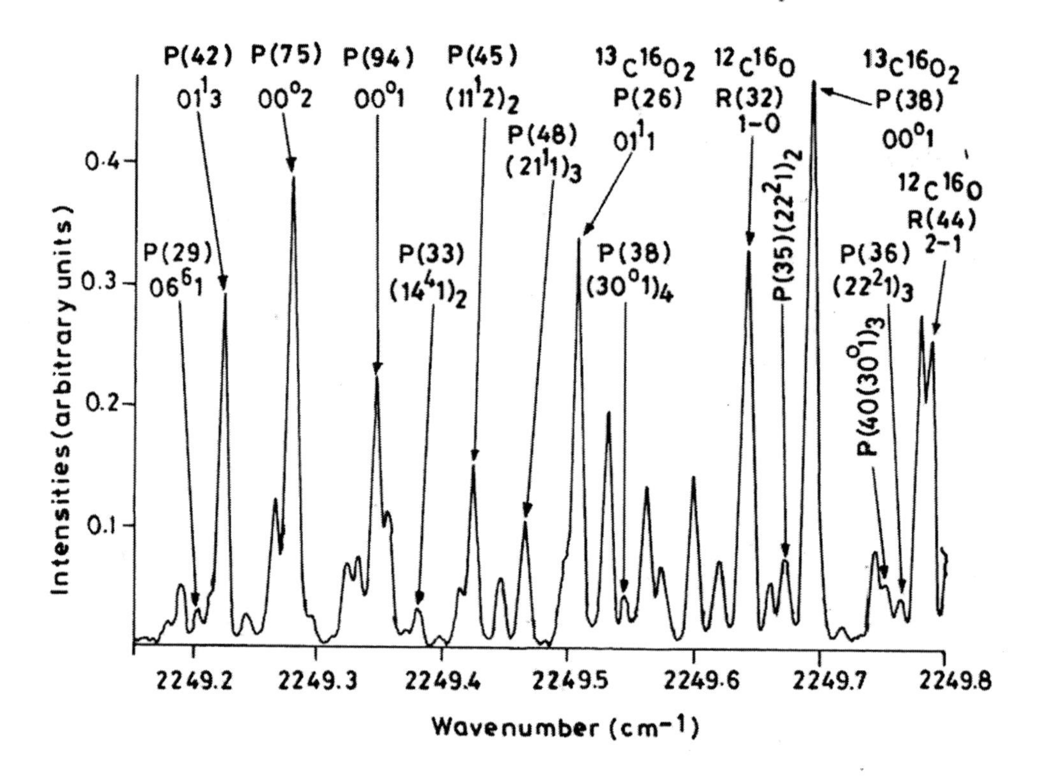

Figure 4.9. A part of the experimental emission spectrum of $^{12}C^{16}O_2$ in the range 2249.15–2249.80 cm^{-1}. The P(J) lines in the $\Delta v_3 = -1$ transitions are indicated alongwith the upper vibrational level denoted by $(v_1v_2^lv_3)_n$ where n indicates the sequence number of a Fermi resonance group. Two lines of $^{13}C^{16}O_2$ and two R(J) lines of $^{12}C^{16}O$ are also shown in the figure. (Reproduced from Bailly [211] with permission from Elsevier Science & Academic Press, Orlando, Florida, USA).

Campargue et al [77] have analyzed the rotational structure of the v_1+5v_3 dyad of $^{12}CO_2$ by recording the $(10^05)_{1,2} \rightarrow (10^04)_{1,2}$ and $(10^06)_{1,2} \rightarrow (10^05)_{1,2}$ emission bands. Using a CH_4+O_2 flame, Bailly et al [81] have recorded the flame spectrum of CO_2 in the 3μm region, and have observed ro–vibrational transitions with $J \approx 100$ in the $(10^01)_{1,2} \rightarrow (00^00)$ band.

(c) Intensities of Ro–Vibrational Lines

A number of papers reporting the line intensities of several bands of carbon dioxide are found in the literature [for e.g. refs 212–215, 136, 216, 217, 137, 218, 56, 35, 26]. The determination of absolute intensities of the bands involves use of equations given in section 3.2.6 of this review. Several authors have also determined the value of squared transition dipole matrix element $|R_{nm}|^2$ and its variation with rotational quantum number, and in some cases, the Herman–Wallis factors. Benner and Rinsland [212] have identified and measured intensities of several P,Q and R-branch lines of the $3\nu_2^3$ "forbidden band" of $^{12}C^{16}O_2$, which appears due to $\Delta l = 2$ Fermi interactions between the upper level and the nearby $3\nu_2^1$ and $\nu_1+\nu_2$ levels. Giver and Chackerian [216] have made laboratory measurements of the intensity and Herman–Wallis parameters of the very weak $(31^10)_{IV}\leftarrow(00^00)$ band at 4416 cm^{-1} which is prominent in the Venus' emission window between 4040 and 4550 cm^{-1}. The CO_2 bands prominent in the Venus emission window centred at 7830 cm^{-1} belong to the vibrational sequence $4\nu_1+\nu_3$ and associated hot bands. Giver et al [218] have reported the line intensities, Herman–Wallis parameters and rotationless band strength for the $(40^01)_{IV}\leftarrow(00^00)$ band at 7460 cm^{-1} and the $(40^01)_{I}\leftarrow(00^00)$ band at 7921 cm^{-1} and also reanalyzed the $(40^01)_{II}\leftarrow(00^00)$ and $(40^01)_{III}\leftarrow(00^00)$ bands. Recently, Teffo et al [26] have measured the line intensities in the 40013←00001, 40014←00001, 10031←00001 and 10032←00001 bands. Both $\nu_1+3\nu_3$ dyad bands and $4\nu_1+\nu_3$ pentad bands belong to the $\Delta P=11$ series. Absolute ro–vibrational intensities of the $(0003)_I\leftarrow(1000)_{II}$ band of $^{12}C^{16}O_2$ centred at 5687.17 cm^{-1} (occurring in the Venus emission window) and its Fermi dyad partner $(00^03)_I\leftarrow(10^00)_I$ band centred at 5584.39 cm^{-1} have been measured by Kshirsagar et al [56].

Mate et al [35] have measured the intensity of the collision induced band at 4680 cm^{-1} produced in the process CO_2 $(\nu_3 = 1)$ + N_2 $(v=1)\leftarrow CO_2$ $(\nu_3=0)+N_2(v=0)$, by using CO_2/N_2 mixtures with a partial density of CO_2 between 2 and 5 amagats, and maximum total density of the mixture upto 10 amagats. The absorption process has been treated as due to an induced dipole moment in a collision of two axially symmetric molecules.

Tashkun et al [219a] have made global treatment of the vibrational–rotational spectral data of $^{13}C^{16}O_2$. In the fittings for the effective dipole moment operator, several line intensity data from FT and few diode laser measurements pertaining to $\Delta P = 1$ and 3 series [table 5, ref 219a] were used. Quite recently similar fittings for $\Delta P = 1$ and 2 series of $^{16}O^{12}C^{17}O$ and $^{16}O^{12}C^{18}O$ have been made by Teffo et al [tables 5 and 7, ref. 219b].

(d) Pressure Broadening, Shift and Line Mixing

Reports on the determination of self broadening coefficients for $^{12}C^{16}O_2$ are: in the 4.3 μm [220], ν_3 and $\nu_1+\nu_3$ bands [221], 10012–10001 band [222], and for $^{13}C^{16}O_2$: in the laser band regions at 883.145, 913.425 and 1017.659 cm^{-1} [223]. N_2 induced broadening coefficients [220, 221] and self induced pressure shift in the 9.4 and 10.4μm bands of CO_2 [224] are also examples of reported work.

Line mixing effects are observed at high pressures (densities of several amagats). Line mixing in the Q–branches of $(11^10)_I$–(02^20) band [225], $(20^00)_I\leftarrow(01^10)$ and $(12^20)_I\leftarrow(01^10)$ bands [226], interbranch line mixing in (10^01) and (02^01) combination bands [227] of CO_2; and Q–branch line mixing in the $(20^00)_I\leftarrow(01^10)$ and $(12^20)_I\leftarrow(01^10)$ bands of CO_2 perturbed by N_2, O_2 and air [228] are to mention some of the reported work.

(B) HCN and its Isotopomers

For HCN, all three vibrational modes ν_1, ν_2 and ν_3 are IR active and a number of papers on the spectra and intensities of the bands of HCN and its isotopomers with D, ^{12}C, ^{13}C, ^{14}N and ^{15}N are reported.

(a) Absorption

Analysis of bands of HCN, and with isotopes of C and N, involving ν_1 and combination bands in the 3300 cm^{-1} region [229], ν_3 and few bands involving few quanta of ν_2 in the 2100 cm^{-1} region [230], overtone bands in the 5400–15100 cm^{-1} region yielding a large set of vibrational (ω,x and y) constants and ro–vibration (α and γ) constants [231], bands in the 14μm region [232], and ν_2 band of HCN [233] are examples of earlier reported work.

Spectra of overtone bands from 4800 to 9600 cm^{-1} involving some new transitions of the bending combination modes of HCN were reported by Quapp et al [234], who have also applied l–type resonance corrections in their analysis. The spectra of $H^{12}C^{14}N$ and $H^{12}C^{15}N$ between 500 and 10,000 cm^{-1} have been analyzed be Maki et al [127] who have (as mentioned in section 3.2.4.3) observed Coriolis interaction involving $\Delta v_3 = -1$, $\Delta v_2 = 3$ and $\Delta l = \pm 1$ for many levels, and have reported the Coriolis interaction constants, l–type resonance constants and a number of vibrational and rotational parameters for the molecules.

(b) Emission

The emission spectra of $H^{12}C^{14}N$, $H^{12}C^{15}N$ and $H^{13}C^{14}N$ in the 400–850 cm^{-1} region involving v_2=1–11, at temperatures of the order of 1370 K have been reported by Maki et al [14]. Quite recently, Quapp et al [235] have reported the $H^{13}C^{15}N$ emission spectrum in the same region (involving v_2 = 1–5). New Coriolis interactions for the three isotopomers were found and analyzed by Maki et al [14] who have reported the ro–vibrational constants for the isotopomers, after correcting for l–type resonances.

The emission spectrum of $D^{13}C^{15}N$ at 1370K in the 450–700 cm^{-1} region has been analyzed by Quapp et al [236], while that of $D^{12}C^{14}N$ and $D^{13}C^{14}N$ at temperatures of 1370K and 1520K respectively have been analyzed by Mollmann et al [16]. Ro–vibrational transitions involving high bending states upto v_2=9 for $D^{13}C^{14}N$ and upto v_2=11 for $D^{12}C^{14}N$ were analyzed by Mollmann et al [16] who have also verified the assignment of the six laser transitions observed in $D^{12}C^{14}N$ by taking into account the Coriolis interaction between the J = 21 levels of the 02^02 and $09^{1e}0$ states.

Table 4.4 lists the parameters and constants for the vibrational term values of $D^{12}C^{14}N$ and $D^{13}C^{15}N$ reported by Mollmann et al [16]. The parameters are found in equation (3.45) of this review and were necessary to accurately fit the observed high resolution data.

Table 4.5 shows the parameters of equation (3.47) used to describe the rotational constants of $H^{12}C^{14}N$ and $H^{13}C^{15}N$. The parameters for $H^{12}C^{14}N$ are reproduced from Maki et al [14] who had reported the constants for $H^{13}C^{15}N$ from some preliminary measurements only. Consequently, the accurate and reliable constants for $H^{13}C^{15}N$ are reproduced from Quapp et al [235].

Table 4.6 lists the parameters for describing the centrifugal distortion constants of $H^{12}C^{14}N$ [from ref. 14] and $H^{13}C^{15}N$ [from ref. 235]. The parameters are found in equations (3.48) and (3.49). It can be seen from table 4.6 that the constant ε_3 for the stretching mode ν_3 could not be determined for both isotopomers and hence it was set to zero [14,235].

Table 4.4 Molecular parameters (in cm–1) for the vibrational energy levels of two isotopomers of DCN (Reproduced from Mollmann et al [16] with permission from Elsevier Science & Academic press, Orlando, Florida, USA).

Parameter	$D^{12}C^{14}N$	$D^{13}C^{15}N$
ω_1	2702.54857(309)[a]	2652.57355(135)
ω_2	579.66395(379)	570.22266(125)
ω_3	1952.28117(773)	1911.01287(58)
x_{11}	–20.16621(148)	[–20.88][b]
x_{22}	–2.14563(108)	–1.989603(581)
x_{33}	–6.96684(377)	[–6.77]
x_{12}	–15.78531(311)	–14.80408(129)
x_{13}	–32.28384(24)	[–28.78]
x_{23}	3.40367(937)	2.589881(529)
g_{22}	3.255652(297)	3.132852(187)
y_{222}	0.025484(73)	0.021575(153)
y_{112}	–0.02486(147)	[–0.023]
y_{122}	0.06415(23)	[0.062]
y_{223}	–0.15357(271)	[–0.15]
y_{233}	–0.12016(449)	[–0.12]
y_{1ll}	–0.04450(44)	[–0.044]
y_{2ll}	–0.017193(81)	–0.013342(92)
y_{3ll}	0.009266(351)	[0.0092]
z_{2222}	–0.0008440(106)	–0.0007401(183)
z_{2233}	0.02008(125)	[0.02008]
z_{22ll}	0.0008724(118)	0.0008048(143)
z_{12ll}	–0.001001(126)	[–0.001]
z_{llll}	–0.0001044(32)	–0.00011737(135)
z_{22222}	0.00000830(56)	0.00000704(81)
z_{222ll}	–0.00001263(89)	–0.00001194(76)
z_{2llll}	0.00000332(40)	[0.0000033]
Std.dev. of fit	1.7[c]	3.6
#Weighted measurements in fit.	47	33

[a]The uncertainty (one standard deviation) in the last digits is given in parentheses.
[b]Values enclosed in square brackets were fixed during the fit.
[c]Since the standard deviations come from weighted fits, they are dimensionless.

Table 4.5 Rotational constants (in 10–3 cm–1) for two isotopomers H12C14N (from ref. 14) and H13C15N (from ref. 235) (Reproduced with permission from Elsevier Science & Academic press, Orlando, Florida, USA).

Parameter	$H^{12}C^{14}N$	$H^{13}C^{15}N$
B_e	1484.775403(1981)[a]	1402.549561(16)[a]
α_1	10.42987(365)	–9.210(11)
α_2	–3.57053(215)	3.042(6)
α_3	10.00542(259)	–9.268(11)
γ_{11}	–0.146922(1875)	[–0.137][b]
γ_{22}	0.046480(908)	0.0407(14)
γ_{33}	–0.029213(328)	[–0.028]
γ_{12}	0.194157(1125)	0.169(11)
γ_{13}	0.193750(3530)	–0.156(21)
γ_{23}	–0.114872(2963)	–0.099(12)
γ_{ll}	–0.195086(633)	–0.1813(8)
γ_{111}	–0.0013230(3205)	[–0.0015]
γ_{222}	0.0003419(1358)	0.00071(21)
γ_{333}	[0.0]	[0.0]
γ_{112}	[0.0]	[0.0]
γ_{113}	0.0027895(11604)	[–0.0025]
γ_{122}	0.0035292(2318)	–0.0035(3)
γ_{223}	–0.0069974(12356)	–0.009(2)
γ_{133}	[0.0]	[0.0]
γ_{233}	[0.0]	[0.0]
γ_{123}	0.0155798(8297)	0.030(21)
γ_{1ll}	–0.0028922(9347)	0.0052(2)
γ_{2ll}	–0.0040592(1315)	–0.0035(2)
γ_{3ll}	0.0083170(8158)	–0.0134(13)
γ_{2222}	0.00005602(602)	0.000014(11)
γ_{2223}	0.0005613(1565)	0.00065(24)
γ_{11ll}	0.001072(282)	[0.001]
γ_{22ll}	–0.00019240(929)	–0.000149(15)
γ_{13ll}	–0.003556(863)	[–0.0039]
γ_{23ll}	–0.0008365(1397)	–0.00148(40)
γ_{llll}	0.00007415(422)	0.000070(5)
Std. dev. of fit	4.3	4.3
#Weighted measurements in fit	105	46

[a]The uncertainty (one standard deviation) in the last digits is given in parentheses.

[b]Values enclosed in square brackets were fixed during the fit.

Table 4.6 Centrifugal distortion constants (in cm–1) for two isotopomers H12C14N (from ref. 14) and H13C15N (from ref. 235). (Reproduced with permission from Elsevier Science & Academic press, Orlando, Florida, USA).

Parameter	$H^{12}C^{14}N$	$H^{13}C^{15}N$
$D_e \times 10^6$	2.85873(141)[a]	2.5625(3)
$\beta_1 \times 10^8$	–3.5641(1250)	–2.99(17)
$\beta_2 \times 10^8$	6.1468(1054)	5.442(79)
$\beta_3 \times 10^8$	0.4510(1713)	0.71(20)
$\beta_{11} \times 10^9$	[0.000][b]	[0.0]
$\beta_{22} \times 10^9$	0.749(123)	–0.145(95)
$\beta_{33} \times 10^9$	[0.000]	[0.0]
$\beta_{12} \times 10^9$	8.693(689)	6.78(45)
$\beta_{13} \times 10^9$	4.300(1423)	5.8(25)
$\beta_{23} \times 10^9$	–0.902(732)	–3.90(49)
$\beta_{ll} \times 10^9$	1.423(433)	–0.58(30)
$\beta_{1ll} \times 10^9$	–2.196(469)	–0.95(19)
$\beta_{2ll} \times 10^9$	–0.581(36)	–0.18(3)
$\beta_{3ll} \times 10^9$	[0.700]	1.32(19)
SD of fit	7.3	3.9
#Weighted measurements	86	44
$H_e \times 10^{12}$	2.356(45)	1.999(59)
$\varepsilon_1 \times 10^{12}$	0.182(43)	0.43(11)
$\varepsilon_2 \times 10^{12}$	0.782(32)	0.63(10)
$\varepsilon_3 \times 10^{12}$	[0.0]	[0.0]
$\varepsilon_{22} \times 10^{12}$	[0.05]	0.23(19)
$\varepsilon_{ll} \times 10^{12}$	–0.288(13)	–0.210(11)
SD of fit	2.0	2.2
#Weighted measurements	31	18

[a] The uncertainty (one standard deviation) in the last digits is given in parentheses.
[b] Values enclosed in square brackets were fixed during the fit.

(C) N_2O and its Isotopic Variants

Even though N_2O has the same number of electrons as CO_2, it is not symmetrical and has the form N–N–O. Among the recently reported papers, mention is made on the measurement of line positions and line strengths of several isotopic species like $^{14}N_2{}^{16}O$, $^{14}N^{15}N^{16}O$, $^{15}N^{14}N^{16}O$, $^{14}N_2{}^{18}O$ and $^{14}N_2{}^{17}O$ by Toth [59] in the 3515–7800 cm^{-1} range in absorption; and the refined investigations in absorption of the overtone spectrum of $^{14}N_2{}^{16}O$ using intracavity laser spectroscopy (ICLAS) and by FTS (5905–7262 cm^{-1}) by Weirauch et al [237], who have analyzed a new hot band at 7229 cm^{-1}. References to earlier work on the molecule are found in the papers of Toth [59] and Weirauch et al [237]. A summary of all available vibrational observations in $^{14}N_2{}^{16}O$ above 6500 cm^{-1} have been given by Weirauch et al [table 3, ref. 237], who have used the cluster labeling notation $[n_r, l_2, i]$, where $n_r = 2v_1 + v_2 + 4v_3$ is the polyad quantum number of the upper state and i is the ordering number within the cluster (increasing with energy). Bailly and Vervloet [80] have analyzed the emission spectrum of $^{14}N_2{}^{16}O$ in the 4.5μm region involving transitions $v_1v_2^{\ell_2}v_3 \rightarrow v_1v_2^{\ell_2}(v_3-1)$ occurring between highly excited vibrational states, by exciting a gas mixture of of N_2O and N_2 with a 13.5 MHz radiofrequency discharge. The addition of N_2 was found to increase the emission of N_2O quite significantly. Bailly and vervloet [80] observed 10 new vibrational transitions between highly excited levels and also detected the previously unreported very weak transitions $00^05 \rightarrow 00^04$ and $04^04 \rightarrow 04^03$.

A number of papers on the measurement of line strengths [for e.g. refs. 238–240] and absolute intensities [for e.g., refs. 241, 242, 27] of N_2O bands have been reported during the last few years. Hartmann et al [243] have investigated the line mixing effects in Q branches of $\Sigma \leftrightarrow \Pi$ bands of N_2O by using three different instrumental set ups including FT measurements in the 17μm region, using N_2 and O_2 as perturbers, at pressures near 0.5 and 1 atm. A theoretical model has been used by Hartmann et al [243] to generate line mixing parameters under atmospheric conditions and comparison between computed stratospheric emissions and the values measured using a balloon–borne high resolution FT instrument was made. The model was found to account correctly for the line mixing effects (as shown in figures 4 and 5 ref. 243).

(D) OCS and OCSe

A number of FT absorption papers have been reported for OCS (for e.g., refs. 244–254, 60]. Fayt et al [255] have made global ro–vibrational analysis for $^{16}O^{12}C^{32}S$ in the ground electronic state using data from FTIR and other techniques, and have calculated ro–vibrational energy levels from a set of effective parameters upto the 8th power in J (J+1). Tolonen et al [247] have analyzed the overtone band $2v_2$ of $^{16}O^{12}C^{32}S$ by taking into account of *l*–type resonance (for $v_2 = 2$, the Σ and Δ_e states perturb each other whereas the Δ_f state remains unperturbed). From the relative intensity measurements of bands in the 2510–3150 cm^{-1} region, Maki et al [248] have reported the relative transition moments. Maki et al [248] observed that due to the resonance coupling of the levels of type (v_1,v_2,v_3, *l*) and (v_1–1, v_2+4, v_3, *l*), there is an enhancement in the intensity of the 04^02–00^01 transition, and found that due to the strong mixing of the wavefunctions of the upper states, the 04^02–00^01 and 10^02–00^01 transitions have equal intensities.

Accurate wavenumbers for the lines of the $2v_2$ band of OCS, which are widely used for calibration purposes, have been reported by Horneman et al [251]. Rbaihi et al [253] have recorded the spectrum of OCS from 4800 to 8000 cm^{-1} with near Doppler resolution and have observed a local perturbation in the 00^03 state. This perturbation is caused by the $1,10^0,0$ state, which crosses the 00^03 state at J = 30. The perturbation could be seen in both P and R branches of the 00^03–00^00 band at J′ = 30. Naim et al [254] have analyzed the OCS spectra between 3700 and 4800 cm^{-1} and found that heavy mixing of states occur through anharmonic resonances, due to which, intensity transfers are observed such that even unobservable bands could be observed. When the states lie close the each other, the intensity transfer was found to be J–dependent. Naim et al [254] have also confirmed the intensity analyses for the bands in the 1800–3100 cm^{-1} region made by earlier workers [252,256].

The OCSe molecule, like OCS, belongs to the linear 16–valence electron triatomic species. The Se substitution shifts down the fundamental modes ν_2 and ν_3. Some examples of the work done on this molecule include study of the $2\nu_1$ band [257], ν_1 and $\nu_1+\nu_3$ band regions [258], ν_2 and ν_3 bands [259], ν_2, $2\nu_2$ and ν_3 regions [260]; and of 18 band systems in the 400–6700 cm^{-1} region comprising of cold, difference and numerous hot bands, giving access to 81 v,*l* vibrational sublevels [261]. Litz et al [261] have compared the anharmonic interaction schemes for OCS and OCSe and found that for both molecules, the same interaction scheme could be applied. Anharmonically interacting polyads and some local crossings with fourth order Coriolis resonances involving levels (v_1,v_2^{l}, v_3) and $(v_1-1, (v_2+3)^{l+1}, v_3+1)$ have been observed for $^{16}O^{12}C\,^{80}Se$ [261].

(E) CS_2

Survey and references to earlier work before 1996 on the high resolution spectrum of CS_2 have been given in the papers of Cheng et al [262] and Blanquet et al [263]. The perturbation between 00^03 and 16^01 levels of $^{12}C^{32}S_2$ [263] which is found to be weak for $^{12}C^{34}S_2$ and occurring at higher J values (J≈90) [264], wavenumbers of the ν_2 band and Fermi resonance involving ν_1 and $2\nu_2$ levels of $^{12}C^{34}S_2$ [262], and the Fermi and *l*–resonances involving $\nu_1 + \nu_2$ and $3\nu_2$ states of $^{12}C^{32}S_2$ [117] are some interesting studies on the molecule. In the $\nu_1+\nu_2$ band, perturbation causes the P-branch to be strikingly stronger than the R-branch, and the lines on the R side nearly disappear around J≈26 [fig. 7, ref. 117]. Due to the very good S/N ratio of their spectrometer, Blanquet et al [265] have observed the very weak 02^03–00^00 transition, and detected a very small perturbation in the $\nu_1+3\nu_3$ band around J≈80, which has been attributed to a resonance between the 10^03 and 18^01 levels.

(F) Other Linear Triatomic Molecules

The high resolution FT spectra of cyanogen halides like FCN [266, 57], ClCN [267–270], BrCN [271, 52] and ICN [272] have been reported in the literature. Farkhsi et al [266] have discussed the anharmonic resonances in FCN, detected a perturbation in the 30^00 vibrational state, and found that the 30^00–$(01^10)_e$ band borrows intensity from the 02^01–$(01^10)_e$ band. The mixing becomes heaviest at J ≈ 45 when both bands appear with similar intensity. Global ro–vibrational analysis of the main isotopomers of FCN has been performed by Farkhsi et al [57]. All available FTIR data and data from other techniques have been used by Saouli et al [270] in their global analysis applied to $^{35}Cl^{12}C^{14}N$ and $^{37}Cl^{12}C^{14}N$, to obtain a set of 80 molecular parameters for the two isotopic species. Burger et al [52] have considered the *l*–type resonance between $(02^00)_e$ and $(02^20)_e$ levels of BrCN, treated the Fermi resonance linking the $(02^00)_e$ and $(10^00)_e$ levels of the molecule (as mentioned in section 3.2.4.1), and have deduced some deperturbed molecular parameters for $^{79}BrCN$ and $^{81}BrCN$.

Absorption spectra of DCP [273, 274] with evidence for Fermi resonance between 10^00 and 00^02 levels [274], ClCP in the 100–4000 cm^{-1} region [275], matrix absorption FT spectrum of the $\nu_1(\sigma)$ (Si–C) stretching mode of linear HCSi in Ar at 10K [276] and the FT emission spectrum of three bands of the $\tilde{A}^2\Sigma^+-\tilde{X}^2\Pi_i$ transition of HCSi radical in the

9000–14000 cm^{-1} region [277] are some examples of additional reported work on linear triatomic molecules.

4.3 FOUR ATOMIC MOLECULES

The high resolution spectra of linear four atomic molecules are of particular interest to both experimentalists and theoreticians, mostly due to the bending state dynamics, anharmonic resonances, and vibrational and rotational *l*–type resonances. In this review acetylene and its isotopomers will be discussed in detail, and other molecules in lesser detail.

(A) Acetylene and its Isotopomers

A large number of research papers on the high resolution FT spectra of acetylene and its isotopomers have been published in the literature. C_2H_2 is one of the molecules listed in the HITRAN database [193]. Some examples of reported work include line positions of bands and molecular constants in $^{12}C_2H_2$ [46, 278–283, 55], $^{13}C_2H_2$ [47, 118, 284–286], $^{13}C_2HD$ [287] and $^{12}C^{13}CH_2$ [279, 284, 288, 289]. Absorption spectra of $^{12}C_2H_2$ have been recorded in the 50–1450 cm^{-1} region with a resolution better than 0.005 cm^{-1} by Kabbadj et al [46]. A complete set of bending energy levels with $\sum_t v_t \leq 4$ (with t = 4,5) including the rotational energy levels from v_2=1 were fitted to 80 molecular parameters. This simultaneous fit involved 43 bands and the use of a full Hamiltonian matrix. The molecular parameters for $^{12}C_2H_2$ reported by Kabbadj et al [46] are presented in table 4.7. The various parameters shown in table 4.7 appear in the expressions for diagonal and off–diagonal elements formulated by Herman et al [114] and were used by Kabbadj et al [46] with certain corrections.

The analyses of anharmonic resonances and *l*–type resonances in acetylene and its isotopomers have been done by several workers. The $\nu_3(\Sigma_u^+)$/ $\nu_2+\nu_4+\nu_5$ (Σ_u^+) Fermi resonance in $^{12}C_2H_2$ has been analyzed by several workers [130, 280–282, 291], in $^{13}C_2H_2$ by Alboni et al [85] and in $^{12}C^{13}CH_2$ by D'Cunha et al [289]. This resonance is strong in $^{12}C_2H_2$ [119] but weak in $^{13}C_2H_2$ [85]. The same resonance becomes strong in $^{13}C_2H_2$ by adding a ν_1 quantum in the dyad $\nu_1+\nu_3(\Sigma_u^+)/\nu_1+\nu_2+\nu_4+\nu_5$ (Σ_u^+) [292]. Keppler et al [290] have analyzed the ν_1+ $\nu_3(\Sigma_u^+)$/ $\nu_1+\nu_2+\nu_4+\nu_5$ (Σ_u^+) Fermi resonance in $^{12}C_2H_2$. The $\nu_3(\Sigma^+)$/ ν_2+ $\nu_4+\nu_5$ (Σ^+) Fermi resonance has been found to be weak in $^{12}C^{13}CH_2$ [289] due to which, the ν_2+ $\nu_4+\nu_5$ (Σ^+) band is only 10% as strong as the $\nu_3(\Sigma^+)$ band, whereas the intensity sharing between the two levels is found to be near total in $^{12}C_2H_2$.

Table 4.7 Molecular Parameters (in cm–1) for 12C2H2 resulting from simultaneous fit of all subbands involving levels up to $\sum_t v_t$ = 4(t=4,5) (Reproduced from Kabbadj et al [46] with permission from Elsevier Science & Academic press, Orlando, Florida, USA).

ω_4^0=609.015766(203)	Y_{444} = 0.0154207(700)	y_5^{44} =0.0650461(453)
ω_5^0=729.1716034(222)	Y_{445} = – 0.021635(254)	r_{45}^0 = – 6.4664304(827)
ω_2^0=1974.3160371(684)	Y_{455} = 0.1565390(909)	r_{445} =0.2075197(388)
x_{44}^0 =3.058842(226)	Y_{555}=0.01399903(577)	r_{455}= – 0.0936683(305)
x_{45}^0 = – 2.433314(174)	y_4^{44} = –1.272(680) x 10^{-4}	r_{45}^J = 1.954565(625) x 10^{-4}
x_{55}^0 = – 2.3357345(192)	y_4^{45} = –0.0221264(988)	z_{4455}=7.2668(403) x 10^{-3}
g_{44}^0 =0.781776(137)	y_4^{55} = –0.0511110(344)	z_{4445}= – 6.8501(779) x 10^{-3}
g_{45}^0 =6.582333(117)	y_5^{55} = –6.94564(872) x 10^{-3}	z_{4555}= – 3.7572(198) x 10^{-3}
g_{55}^0 = 3.4899715(184)	y_5^{45} =0.014461(104)	z_{45}^{45} = 0.0300235(819)
B_0 = 1.176646179(166)	γ^{44} = –6.2940(689) x 10^{-5}	γ_4^{44} = –1.384(340) x 10^{-6}
B_2 = 1.17046497(109)	γ^{45} = –2.15196(210) x 10^{-4}	γ_4^{45} = 1.115(124) x 10^{-6}
α_4= –1.356484(402) x 10^{-3}	γ^{55}= –1.0674981(898) x 10^{-4}	γ_4^{55} = 9.09(128) x 10^{-7}
α_5 = –2.229220(113) x 10^{-3}	γ_{444} = 2.850(162) x 10^{-6}	γ_5^{44} = –2.055(289) x 10^{-6}
γ_{44} = –6.109(205) x 10^{-6}	γ_{445} = –4.810(149) x 10^{-6}	γ_5^{45} = –1.1676(145) x 10^{-5}
γ_{45} = –1.8963(228) x 10^{-5}	γ_{455} = 1.0545(716) x 10^{-6}	γ_5^{55} = –1.6147(420) x 10^{-6}
γ_{55} = 2.03922(894) x 10^{-5}	γ_{555} = –1.602(263) x 10^{-7}	
D_0 = 1.626536(236) x 10^{-6}	β_{44} = 4.561(225) x 10^{-9}	β^{44} = –1.2657(573) x 10^{-8}
D_2 = 1.61982(401) x 10^{-6}	β_{45} = –1.837(573) x 10^{-10}	β^{45} = 8.26(109) x 10^{-10}
β_4 = 4.1036(361) x 10^{-8}	β_{55} = –1.0011(583) x 10^{-9}	β^{55} = 1.401(158) x 10^{-9}
β_5 = 2.4559(132) x 10^{-8}		
H_0 = 1.3304(943) x 10^{-12}	H_5 = 2.033(276) x 10^{-13}	H_2 = –8.75(392) x 10^{-12}
q_4^0= 5.252116(273) x 10^{-3}	q_{54} = 1.01355(238) x 10^{-4}	q_4^J = –3.7024(163) x 10^{-8}
q_5^0= 4.6637801(800) x 10^{-3}	q_{55} = 3.46972(644) x 10^{-5}	q_5^J = –3.86490(851) x 10^{-8}
q_{44} = – 2.1866(262) x 10^{-5}	q_4^k= 9.531(174) x 10^{-6}	q_4^{JJ} = –2.142(101) x 10^{-12}
q_{45} = 7.4666(254) x 10^{-5}	q_5^k= 7.039(357) x 10^{-6}	q_5^{JJ} = 1.114(419) x 10^{-13}
ρ_4^0= 5.592(298) x 10^{-8}	ρ_{45} = –1.8273(308) x 10^{-8}	ρ_5^4= 9.642(851) x 10^{-9}
ρ_5^0= 1.750(218) x 10^{-8}	ρ_{55} = –1.3795(977) x 10^{-8}	ρ_{44}= –1.2573(966) x 10^{-8}
ρ_4^5= 1.6254(729) x 10^{-8}	ρ_5^J = –1.060(170) x 10^{-12}	

Number of fitted data = 2307; R.M.S. error = 0.000439 cm^{-1}
Estimated errors (1σ) are given in parentheses in units of the last figure quoted.

Several anharmonic resonances in $^{13}C_2H_2$ have been analyzed by Venuti et al [292]. Smith and Winn [125] have analyzed several types of Fermi resonances via coupling constants K_{1234} and K_{2345}, and also analyzed several Darling–Dennison resonances in the C–H overtone spectrum of $^{12}C_2H_2$ below 10,000 cm^{-1} [table II, ref. 125]. As mentioned in section 3.2.4.1, the strong Fermi resonance $\nu_1(\Sigma_g^+)/\nu_2+2\nu_5(\Sigma_g^+)$ in $^{13}C_2H_2$ has been studied by Di Lonardo et al [118], by detecting the hot bands $\nu_1(\Sigma_g^+)\leftarrow\nu_5(\Pi_u)$ and $\nu_2+2\nu_5$ $(\Sigma_g^+, \Delta_g)\leftarrow\nu_5(\Pi_u)$ in the 4μm region. Lievin et al [293] have recorded the FT absorption spectrum of $^{12}C_2HD$ from 4600 to 9000 cm^{-1}. By gathering all spectroscopic data on the molecule from 500 to 16000 cm^{-1}, they have identified several anharmonic resonances in the molecule [table 5, ref. 293] and were able to obtain information on the dynamics and on the evolution of the absorption intensity among overtones.

l–type resonances (vibrational and rotational) have been studied in $^{12}C_2H_2$ by several workers [114, 130, 280, 282, 283] and in $^{13}C_2H_2$ by Di Lonardo et al [47] and by Venuti et al [292]. Weber et al [294] have observed intensity perturbations due to *l*–type resonances in six hot bands of $^{12}C_2H_2$ in the 13.7μm region involving $\nu_4+\nu_5\leftarrow\nu_4$ and $2\nu_5\leftarrow\nu_5$ transitions, but could not observe any intensity perturbation due to *l*–resonance in the ν_5 fundamental band. Vander Auwera [25] measured the absolute intensities of the $\nu_4+\nu_5(\Sigma_u^+)\leftarrow GS(\Sigma_g^+)$ band and the forbidden $\nu_4+\nu_5(\Delta_u)\leftarrow GS(\Sigma_g^+)$ band in $^{12}C_2H_2$, observed near 1328.081 and 1342.821 cm^{-1} respectively. Both bands exhibited significant Herman–Wallis dependence of line intensities, which were explained due to *l*–type resonance and Coriolis coupling [figures 1 and 2, ref. 25]. The intensity asymmetry between P and R branches observed in the $\nu_4+\nu_5(\Sigma_u^+)\leftarrow GS(\Sigma_g^+)$ band [fig. 1, ref. 25] has been interpreted as arising from intensity borrowing from $\Delta l=\pm1$ transitions, induced by Coriolis mixing of the levels.

Quite recently, Di Lonardo et al [295] have studied the FT absorption spectrum of isotopically enriched $^{12}C^{13}CH_2$ (isotopic purity 99%) in the 450–3200 cm^{-1} region, with an effective resolution ranging from 0.004 to 0.006 cm^{-1}. 53 bands of the isotopomer were analyzed involving bending states upto$v_t=v_4+v_5=4$. Bands involving states upto $v_t=3$ were analyzed simultaneously by a model Hamiltonian, taking into account the vibrational and rotational *l*–type resonances, analogous to that used in the study of the bending states in $^{12}C_2H_2$ [46], and in $^{13}C_2H_2$ [47], and based on the formulation of Herman et al [114]. A set of 71 statistically significant molecular parameters for $^{12}C^{13}CH_2$ have been reported by Di Lonardo et al [295], who have also given a detailed listing of the diagonal and off–diagonal contributions to the Hamiltonian matrix.

A very interesting observation of lack of intensity alternation was made by Herman et al [296] in the R (J) lines with J = 20, 21 and 22 in the emission spectrum of $^{12}C_2H_2$ (vibrationally excited in a radiofrequency discharge) involving the ν_3 and $\nu_2+\nu_4+\nu_5$ cold bands. Such a lack of intensity alternation was not observed in the absorption bands. The lines with higher J values were found to have the usual intensity alternation, in the emission spectrum. The lack of intensity alternation for specific lines in the cold bands in emission, has been attributed to the successive emission–absorption processes occurring in discharges [296].

A systematic modelling of the vibration–rotation pattern of energy levels in the ground electronic state of acetylene has been done in a series of papers [297–301] by Professor M. Herman and his collaborators, who have developed the "cluster model" based on the concept of polyads. Each cluster is characterized by new quantum numbers $N_s = v_1+v_2+v_3$, $N_r = 5v_1+3v_2+5v_3+v_4+v_5$, and $k=l_4+l_5$, where N_s is essentially a polyad quantum number pertaining to the total number of stretching quanta, and N_r is another polyad quantum number which depends on the ratio of vibrational frequencies in $^{12}C_2H_2$ given approximately as $v_1:v_2:v_3:v_4:v_5 = 5:3:5:1:1$. Each cluster is labeled by $[N_s, N_r, k, u/g]$ symbol, and for Σ species g^+, g^-, u^+ or u^- are included depending on the behaviour of the vibrational species to reflection, under $\infty\sigma_v$ operation for $D_{\infty h}$ group.

El Idrissi et al [300] have gathered data on 253 vibrational levels in the ground electronic state of $^{12}C_2H_2$ from different spectroscopic investigations on the molecule, upto energy range of 18915 cm^{-1}, and performed a global fit using the cluster model. A total of 29 vibrational constants for $^{12}C_2H_2$ resulting from a fit of the vibrational origins of 219 levels covering spectral range upto 17880 cm^{-1}, have been reported by El Idrissi et al [300]. In a similar way, Di Lonardo et al [301] have gathered data on a total of 134 vibrational levels in the ground electronic state of $^{13}C_2H_2$ covering the range upto 23670 cm^{-1}, and performed a simultaneous fit on 118 bands observed below 13000 cm^{-1}, using the cluster model. Table 4.8 presents the 29 vibrational constants including off–diagonal parameters $K_{3/245}$, $K_{1/244}$, $K_{1/255}$, $K_{11/33}$, $K_{14/35}$ and r_{45} for $^{13}C_2H_2$ ($\tilde{X}\,^1\Sigma_g^+$) state reported by Di Lonardo et al [301], which gives the most accurate set of parameters available for the isotopomer in the literature till now.

Table 4.8 Vibrational constants (in cm–1) determined for 13C2H2 ($\tilde{X}\,^1\Sigma_g^+$) obtained from the global fit to data from 118 levels observed below 13,000 cm–1, using the cluster model. The uncertainties (1σ) are given in parentheses, in units of the last quoted digit. (Reproduced from G.Di Lonardo et al [301] with permission from Professor M. Herman (Bruxelles, Belgium) and from American Institute of Physics, New Work, USA).

Dunham parameters			
$\omega_1^0 = 3374.90(49)$	$x_{11} = -25.43(27)$	$x_{12} = -10.55(25)$	$x_{24} = -11.86(10)$
$\omega_2^0 = 1918.04(31)$	$x_{22} = -7.25(16)$	$x_{13} = -104.26(27)$	$x_{25} = -2.088(80)$
$\omega_3^0 = 3305.55(36)$	$x_{33} = -26.68(19)$	$x_{14} = -13.52(20)$	$x_{34} = -10.02(12)$
$\omega_4^0 = 599.92(19)$	$x_{44} = 3.084(52)$	$x_{15} = -9.88(21)$	$x_{35} = -8.19(16)$
$\omega_5^0 = 727.23(19)$	$x_{55} = -2.267(50)$	$x_{23} = -4.84(18)$	$x_{45} = -2.120(92)$
	$g_{44} = 0.800(48)$	$g_{55} = 3.414(45)$	$g_{45} = 6.625(82)$
Interaction parameters			
$r_{45} = -5.965(71)$	$K_{1/244} = 15.10(65)$	$K_{1/255} = 6.34(14)$	
$K_{3/245} = -15.21(38)$	$K_{14/35} = 10.99(52)$	$K_{11/33} = -103.35(40)$	

The vibrational energy pattern in $^{12}C_2D_2$ ($\tilde{X}\,^1\Sigma_g^+$) state has been studied by Herman et al [123] by using FT absorption data between 5150 and 8000 cm^{-1}, ICLAS data between 12800 and 16600 cm^{-1}, and experimental data from other techniques (references to earlier work on C_2D_2 are also given in ref. 123). A global fit to 88 bands yielded 21 molecular parameters. Vibrational clusters were defined for $^{12}C_2D_2$ which were characterized by only two quantum numbers N_s and k.

Recently, the inter– and intrapolyad structures have been investigated in the vibrational energy pattern of $^{12}C_2H_2$ by Zhilinskii et al [302] who have highlighted a very regular structure in the energy pattern of the (N_r, N_s) sub–polyads, which has been attributed to the existence of two constants of motion. In addition, the presence of both oscillatory and regular contributions to the density of states of the same symmetry has also been demonstrated [302].

(B) HCCX and DCCX (X=F, Cl, Br, I)

FT spectra of HCCF in the 1700–7500 cm^{-1} region have been analyzed by Holland et al [112] who have observed 130 bands in this region, and reported three Fermi resonances $\nu_1/\nu_2+\nu_3$, $\nu_2/2\nu_3$ and $\nu_3/2\nu_4$ in the molecule. Three weak perturbations due to resonances between ν_2 ($\sum^+$) and $\nu_3+2\nu_4$ (Δ) states, between $2\nu_3$ ($\sum^+$) and $3\nu_4+\nu_5$ ($\sum^+$, Δ) states, and between $2\nu_2$ ($\sum^+$) and $\nu_1+2\nu_4$ ($\sum^+$) states, in the overtone spectrum of HCCF, have been analyzed by Holland et al [303] around the 2000 cm^{-1} region, where the density of states is low. Due to the increasing density of states at higher energy, analysis of small perturbations become difficult. Borro et al [304] have included vibrational *l*–resonance, three Fermi resonances and one quartic anharmonic resonance (involving quartic coefficient K_{1244}) in their theoretical model for HCCF and also found that only the $\nu_2/2\nu_3$ Fermi resonance is present in DCCF.

The FT spectrum of HCCCl has been analyzed by McNaughton and Shallard [305] in the 20–1850 cm^{-1} and over 10,000 vibrational–rotational lines were assigned. Borro et al [304] have found only effects of vibrational *l*–resonance and the quartic anharmonic resonance (involving K_{1244}) for HCCCl. In DCCCl, none of the above resonances were found to be important but a new quartic resonance was observed between ν_1 and $\nu_2+2\nu_5$. Wang et al [53] have investigated the ν_1 and ν_2 fundamental bands of DCCCl and found that both bands were heavily perturbed. The ν_1 state is in resonance with the *l*=0 substate of the $\nu_3+4\nu_4$ state, and the ν_2 state is in resonance with the *l*=0 substate of the $\nu_3+4\nu_5$ state. Wang et al [53] have analyzed these resonances and have reported the anharmonic potential constants k_{134444} and k_{235555} to be 0.57 cm^{-1} and 0.44 cm^{-1} respectively for $DCC^{35}Cl$.

Vaittinen et al [128] have analyzed 23 bands of HCCBr including the ν_3, ν_4 and ν_5 bands, and treated the Fermi resonance between ν_3 and $2\nu_5$, and between their higher overtones and combinations. Coriolis and rotational *l*–resonances in some bands were also analyzed. High resolution spectrum of the ν_2 band and some hot bands in the 1750–2450 cm^{-1} region has been analyzed by Halonen [306], who observed weak localized rotational perturbations in some bands. Vaittinen et al [126] have noted that only states with the same symmetry and with the same polyad quantum number $V=11\nu_1+7\nu_2+2\nu_3+2\nu_4+\nu_5$ (harmonic frequencies given in table VI, ref 126) are coupled with each other and hence the Hamiltonian matrix could be factorized to a large number of smaller matrices. However, this model would produce matrices whose size increases rapidly with increasing energy. For the V = 11 polyad, for example, the number of interacting states would be 113. Hence to limit the size of the matrix, only the closest interacting states were accounted by Vaittinen et al [126]. They also treated the $\nu_3/2\nu_5$ Fermi resonance, $\nu_1/\nu_2+2\nu_4$ Darling–Dennison resonance, vibrational l–resonance (r_{45} parameter), and found that in HCCBr, the $2\nu_1$, $\nu_1+\nu_2+2\nu_4$ and $2\nu_2+4\nu_4$ states form a Darling–Dennison resonance triad. Brotherus et al [307] have analyzed 124 bands of DCCBr the 240–990 cm^{-1} region including the band systems ν_5, ν_4, ν_3, $\nu_4+\nu_5$, and $2\nu_4$. The states in rotational l–resonance (footnotes under table 7, ref 307), the Fermi resonance $\nu_3/2\nu_5$ and vibrational l– resonance (r_{45} parameter) were treated by Brotherus et al [307].

Tolonen et al [308] have recorded the FT absorption spectrum of HCCI in certain selected wavenumber ranges between 200 and 3500 cm^{-1}, and have analyzed the five fundamental bands ν_1 to ν_5. Various Fermi and l–resonances have been treated by Ahonen et al [131] in the analysis of the spectrum of HCCI between 2000 and 3000 cm^{-1}. Sarkkinen et al [132] have studied the spectra of DCCI in the 220–550 cm^{-1} region, analyzed the Fermi resonances $\nu_3/2\nu_5$ and $\nu_3+\nu_5/3\nu_5$, and considered various l–resonances in the overtone levels. Recently, Sarkkinen [115] has analyzed Coriolis, Fermi, and several l– type resonances in HCCI and DCCI. They have highlighted the benefits of the simultaneous analysis of many ro–vibrational energy levels for each isotopomer (as done by them), which are for example; revealing weak resonances, determination of resonance parameters (as in the case of DCCI), and smaller error limits in the resulting molecular constants.

(C) NCCN and its Isotopomers

The cyanogen molecule (NCCN) has, like acetylene, three stretching modes $\nu_1(\Sigma_g^+)$ (symmetric CN stretching), $\nu_2(\Sigma_g^+)$ (CC stretching) and $\nu_3(\Sigma_u^+)$ (asymmetric CN stretching), and two doubly degenerate bending modes $\nu_4(\Pi_g)$ and ν_5 (Π_u) respectively. The ν_3 and ν_5 fundamental modes are IR active. Like in acetylene, the vibrational levels are indicated by $(v_1 v_2 v_3 v_4^{\ell_4} v_5^{\ell_5})$ with k= l_4+l_5.

Grecu et al [309] have recorded the high resolution spectrum of $^{14}N^{12}C^{12}C^{14}N$ in the 180–280 cm^{-1} region with a resolution of 0.0018 cm^{-1}. Besides the ν_5 band system, the transitions involving the ν_5 manifold (16 subbands) with v_5 upto 4, were also assigned. Due to the small rotational constant (B_0 = 0.15708769 (14) cm^{-1}), the ν_5 band showed very high line density (upto more than 200 lines per cm^{-1}). In a following paper, Grecu et al [139] reported the absolute ro–vibrational line intensities in the ν_5 fundamental band and in the first hot band. Grecu et al [139] also measured nitrogen broadened spectra and obtained a value of 0.12 cm^{-1}/atm for the nitrogen broadening coefficient for the ν_5 band of $^{14}N^{12}C^{12}C^{14}N$.

Quapp et al [310] have measured the spectrum of $^{15}N^{13}C^{13}C^{15}N$ in the 1950–2150 cm^{-1}. The ν_3 fundamental band, difference band $\nu_1-\nu_5$, and hot band transitions arising from ν_5,$2\nu_5$, $3\nu_5$ and ν_4 states were also analyzed. Only in the case of the $\nu_3+2\nu_5\leftarrow 2\nu_5$ and $\nu_3+3\nu_5\leftarrow 3\nu_5$ transitions, it was found necessary to make l–resonance analysis.

Maki and Klee [133] have made a systematic analysis of the perturbations involving ν_1 mode of NCCN. The levels 1000^00^0, 1000^01^1, $1000^02^{0,2}$ and $1000^03^{1,3}$ showed pronounced perturbations in $^{14}N^{12}C^{12}C^{14}N$, $^{14}N^{13}C^{13}C^{14}N$ and in $^{15}N^{12}C^{12}C^{15}N$. The measurements made on the hot band transitions 1000^01^1—0000^02^0 and 1000^01^1—0000^02^2 near 2100 cm^{-1}, and transition from the ground state 1000^01^1—0000^00^0 near 2560 cm^{-1} involving the 1000^01^1 state of the $^{14}N^{12}C^{12}C^{14}N$ isotopomer showed effects of several perturbations, two each for *e* and *f* levels.

Figure 4.10 shows the Q–branch region of the $\nu_1+\nu_5$ band of $^{14}N^{12}C^{12}C^{14}N$, reported by Maki and Klee [133]. Due to perturbation, the Q (27) line is displaced to higher wavenumbers and the Q (28) line is displaced to lower wavenumbers. This displacement of the Q lines created a very prominent hole in the absorption spectrum, causing a transmission peak at 2561.23 cm^{-1}. The Q–branch transitions shown in fig 4.10 involve the *f* levels of the $\nu_1+\nu_5$ band. Weak perturbation of the *e* levels were also observed in the same band. Maki and Klee [133] have shown that the low J perturbations of the *e* and *f* levels of the 1000^01^1 state in $^{14}N^{12}C^{12}C^{14}N$ are caused by the *e* and *f* levels of the 0102^03^3 state (which is clear from fig. 3, ref. 133). Strong perturbations at higher J values is caused due to the 0102^03^1 state. Other perturbations observed in the spectra recorded in the 1900–2750 cm^{-1} region have also been explained by Maki and Klee [133].

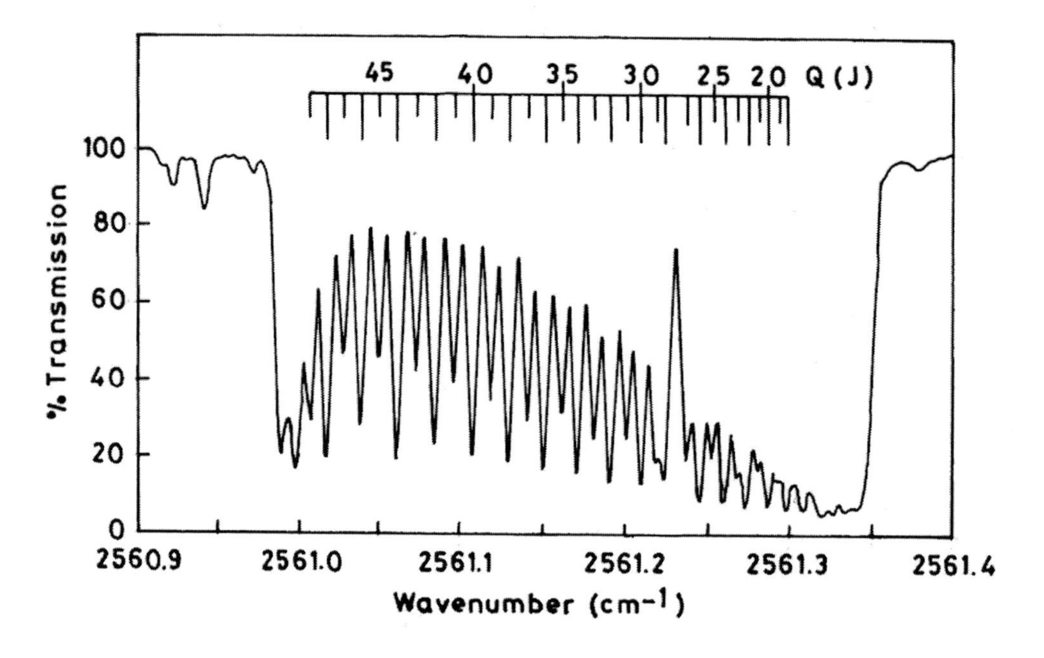

Figure 4.10. The Q–branch region of the $\nu_1+\nu_5$ absorption band of $^{14}N^{12}C^{12}C^{14}N$. Due to perturbation, the Q(27) line is displaced to higher wavenumbers and the Q(28) line is displaced to lower wavenumbers, which has caused a transmission maximum at 2561.23 cm^{-1}. (Reproduced from Maki and Klee [133] with permission from Elsevier Science & Academic Press, Orlando, Florida, USA).

Quite recently, Hochlaf [311] has investigated the six–dimensional potential energy surface of $^{14}N^{12}C^{12}C^{14}N$ by using the "coupled cluster single and double approach with perturbative treatment of triple excitations" CCSD (T) method. The calculated values of the energy levels 0000^0v_5 upto v_5=4 by Hochlaf [311] are in good agreement with the precisely measured FT experimental values of Grecu et al [309].

(D) Other Four Atomic Molecules

The FT spectrum of isocyanogen (CNCN) has been investigated by some workers. Siebert et al [312] has given a survey on the experimental and theoretical work done on the molecule, since the first identification of the molecule in 1989. In the spectrum of $C^{15}NC^{15}N$ recorded in the 1670–2450 cm^{-1} region, Siebert et al [312] have analyzed the ν_1 band system involving 15 subbands, ν_2 band system involving 16 subbands, as well as the upper component of the Fermi triad involving $4\nu_4$, $2\nu_4+\nu_3$ and $2\nu_3$ around 1940 cm^{-1}. The rotational, centrifugal distortion and the *l*–doubling constants were found to depend to a first approximation, linearly on vibrational quantum number v_5. Stroh et al [313] have analyzed the ν_5, ν_4 and $\nu_4+\nu_5$ band systems and also the Fermi resonance between ν_3 and $2\nu_4$ in $^{12}C^{14}N^{12}C^{14}N$. The ordering of the vibrational term values of CNCN was found to compare with that for NCCN.

The FT spectra of the short lived linear difluoroethyne molecule (FCCF) has been reported by some workers [314,315]. The ro–vibrational energy levels, nuclear spin statistics and the five fundamental vibrations of the molecule are analogous to those in acetylene. In the spectrum of $^{12}C_2F_2$ in the 1350 cm^{-1} region, Burger and Sommer [314] have detected and analyzed the ν_3, $(\nu_3 + \nu_4)\leftarrow\nu_4$ and $(\nu_3+\nu_5)\leftarrow\nu_5$ bands while in the 2150 cm^{-1} region, they analyzed the $\nu_2+\nu_3$, hot bands with ν_4 and ν_5, and difference band $\nu_1-\nu_5$. In addition, the anharmonic interaction of ν_3 with $(\nu_2+\nu_4+\nu_5)^0$ was also analyzed. Burger et al [315] have analyzed the ν_3 and $\nu_2+\nu_3$ bands of $^{12}C^{13}CF_2$ of which, the former band was found to be slightly perturbed and the latter strongly perturbed.

4.4. Five Atomic Molecules

Compared to linear triatomic and four atomic molecules, less number of reported work exist on the FT spectra of five atomic linear molecules.

(A) HCCNC and DCCNC

Burger et al [316] have recorded the FT absorption spectra of isocyanoacetylene (HCCNC) in the 350–5000 cm^{-1} region and of DCCNC in the 350–3500 cm^{-1} region. Both HCCNC and DCCNC are linear molecules which have four stretching vibrations ν_1 to ν_4 (all Σ^+ species) and three bending vibrations ν_5, ν_6 and ν_7 (all Π species). Burger et al [316] have recommended the rounded wavenumbers (in cm^{-1}) for the vibrational fundamentals of HCCNC and DCCNC as ν_1= 3339 (2618), ν_2 = 2219 (2158), ν_3 = 2037 (1965), ν_4=955 (929), ν_5=621(480), ν_6=430(425) and ν_7=207(200), where the wavenumbers in parentheses are for DCCNC. The fundamentals ν_1 to ν_5, several combination bands and overtones were observed by Burger et al [316]. The v_5 = 1 state of HCCNC and v_5 = v_7 =1 state of DCCNC were found to be moderately perturbed, while the v_5=v_7=1 state of HCCNC appeared to be strongly perturbed. The anharmonic interactions between ν_4, $2\nu_5^0$ and $2\nu_6^0$ were also analyzed for both molecules [316].

Vigouroux et al [116] have made global ro–vibrational analysis of HCCNC and have obtained a set of 155 molecular parameters (table 3, ref. 116) for the molecule. In the 600–2000 cm^{-1} region, they assigned 53 new IR bands. The polyads upto 2030 cm^{-1} have been described and analyzed in detail. The anharmonic interaction schemes for HCCNC and HCCCN have also been compared by Vigouroux et al [116].

(B) HCCCN and DCCCN

Like HCCNC and DCCNC, the HCCCN and DCCCN molecules have four stretching vibrations ν_1 to ν_4 (all $\sum^+$ species) and three bending vibrations ν_5, ν_6 and ν_7 (all Π species). The rounded wavenumbers (in cm^{-1}) for HCCCN and DCCCN are (from table II, ref. 316), ν_1=3327 (2609), ν_2=2274 (2252), ν_3=2079 (1968), ν_4=864(850), ν_5=663(523), ν_6=499(494) and ν_7=222(212) where the wavenumbers in parentheses are for DCCCN.

The FT spectrum of cyanoacetylene (HCCCN) in the 300–3500 cm^{-1} region has been analyzed by Khlifi et al [317], and the ν_1, ν_2, ν_3, $2\nu_5$, $\nu_5+3\nu_7$, $2\nu_6$, $\nu_5+\nu_7$, $\nu_6+\nu_7$, ν_5, $3\nu_7$ and ν_6 bands have been analyzed and their absolute intensities determined. In a following paper, Arie et al [318] have analyzed 22 bands including the ν_5,ν_6 fundamentals, and associated hot bands in the 450–730 cm^{-1} region. For the ν_5 fundamental band, ro–vibrational lines were assigned from P(100) to R(100), while for the ν_6 band, upto J = 87 were assigned. Arie et al [318] represented the levels with $(\nu_5,\nu_6,\nu_7)^l$ with $l=l_5+l_6+l_7$, as other vibrational quantum numbers were zero. Some bands showed level crossings especially belonging to the $(1,0,1)^l$ levels system with l=0 or 2 (fig. 2, ref. 318), and $(1,0,2)^l$ level system with l = 1 or 3 (fig. 3, ref. 318). Winther et al [121] have observed the ν_1 band system of HCCCN along with several hot bands from ν_5=1, ν_6=1, ν_7=1–4, $\nu_6=\nu_7$=1, ν_5=1, and ν_7=2. From the analysis of the ν_1 band and its associated hot bands, Winther et al [121] observed and analyzed many local perturbations. As mentioned in section 3.2.4.1, HCCCN is known to exhibit resonance between ν_5 and $3\nu_7$ (anharmonic and l–type interactions). Four–state interactions involving ν_4, $\nu_5+\nu_7$, $2\nu_6$ and $4\nu_7$ (rotational and vibrational l–type resonances, and anharmonic interactions) from the analysis of rotational spectra in some excited vibrational states have been studied by Yamada and Creswell [319].

Coveliers et al [122] have recorded the FT spectra of DCCCN in the 200–365 cm^{-1} region and analyzed the $\nu_7\leftarrow$GS, $\nu_6\leftarrow\nu_7$, $\nu_5\leftarrow\nu_7$, $\nu_4\leftarrow\nu_6$ bands, and some of their associated hot bands. In their global analysis on the ν_4,ν_5,ν_6 and ν_7 modes of DCCCN, Coveliers et al [122] used data from other techniques, included vibrational and rotational l–type resonances, considered first order anharmonic resonance between ν_4 and $2\nu_6$, third order anharmonic resonance between ν_4 and $4\nu_7$ (as mentioned in section 3.2.4.1), and introduced the interaction between $2\nu_5^2$ and $2\nu_6^2$. In DCCCN, the resonance system (ν_4,$2\nu_6$, $4\nu_7$) had been analyzed earlier from the millimeter–wave spectra by Plummer et al [320].

The temperature dependent integrated band intensities for the ν_1,ν_2,ν_5 and ν_6 bands of HCCCN have been measured by Khlifi et al [321] in the range 225K $\leq$ T $\leq$ 325K. A decrease (~20%) of integrated band intensities of ν_1,ν_2 and ν_6 bands were observed, while a small increase (~5%) was observed for the ν_5 band, in the studied range of temperatures.

(C) OCCCO and SCCCS

These molecules have no permanent electric dipole moment and hence their pure rotational spectra cannot be observed. The information on rotational and centrifugal distortion constants etc. are obtained only from the analysis of their vibration–rotation spectra. There are indications that both molecules may be constituents of interstellar clouds.

While a number of FT absorption measurements have been made for carbon suboxide (OCCCO) [322–327, 329, 330], there has been less amount of reported work on carbon subsulfide (SCCCS) [for e.g. ref. 135]. The effeective bending potential associated with ν_7 in different states of OCCCO has been calculated by some workers [328, 331, 332] using data obtained from FT experiments. While the OCCCO molecule is considered to have a quasi–linear equilibrium geometry, the SCCCS molecule has a confirmed linear equilibrium geometry. Both molecules have seven fundamental modes of vibration, having vibrational frequencies (in cm^{-1}) [from table 1, ref. 135]: $\nu_1(\sum_g^+)$ = 2196.93 (1663), $\nu_2(\sum_g^+)$ = 787.72 (489.94), $\nu_3(\sum_u^+)$: 2289.80 (2100.10), $\nu_4(\sum_u^+)$ = 1587.39 (1030.18), $\nu_5(\Pi_g)$ = 580.2 (470), $\nu_6(\Pi_u)$ = 540.24 (502) and ν_7 (Π_u) = 18.2558 (93.65) where the wavenumbers in parentheses are for SCCCS. The normal modes of vibration of OCCCO have been drawn in the paper of Fusina et al [326].

In the spectra of C_3O_2, Manz et al [322] have observed Coriolis resonances at J = 24 in the $2\nu_2+\nu_4$ band, at J = 10 in the $\nu_4+(4\nu_6)^0$ band, at J = 28 and 40 in the $\nu_4+(2\nu_5+2\nu_6)^0$ band, at J = 62 in the $\nu_3+(4\nu_6)^0$ band, and at J = 44 in the $\nu_1+\nu_3+(2\nu_7)^0$ band. Weber et al [323] have used FT experiments to record the ν_4 band of C_3O_2 at a resolution of 0.063 cm^{-1} and further used a tunable diode laser to study the ν_4 band. Weber et al [325] observed that in the ν_6 band, the value of B′–B″ was negative, indicating an increased moment of inertia of the molecule in the ν_6 state.

Detailed investigation of the FT spectrum of C_3O_2 recorded in the 1800–2600 cm^{-1} region at a resolution of 0.003 cm^{-1} was made by Fusina et al [326] who analyzed 28 bands associated with ν_3, 7 bands associated with $\nu_2+\nu_4$, and the band $\nu_2+\nu_5+\nu_6$. A number of bands showed rotational perturbation in the upper state. Two strongest perturbations were observed in the $v_7^{\ell_7} = 2^0$ and 5^{1e} vibrational states of the ν_3 manifold, at J = 27 and at J = 48 respectively. Fusina et al [326] considered that these perturbations must be due to high order terms in the Hamiltonian, since the interaction term increases to high powers of J (probably J^6 or J^8 in the two cases of strong perturbation shown in the plots of Δ vs J where Δ = (observed)–(calculated) vibration rotation energy). Fusina et al [331] have reviewed the vibrational energy levels and rotational constants of C_3O_2 for the bending mode ν_7 (whose wavenumber is only 18.3 cm^{-1}) in the ground state manifold, and in the ν_2,ν_3,ν_4, and $\nu_2+\nu_4$ state manifolds. The effective bending potential associated with ν_7 for each of the five states were determined. Weber [332] used the rigid–bender model to treat the large amplitude, low frequency bending vibration ν_7 for C_3O_2. The ν_7 potential function was determined for the states in which ν_2, ν_3, ν_4, ν_6, $2\nu_6$, $\nu_1+\nu_3$, $\nu_1+\nu_4$, $\nu_2+\nu_3$ and $2\nu_2+\nu_4$ were excited. Weber [332] found that in the ground state, the ν_7 potential has a 29 cm^{-1} barrier at the linear position. The barrier increases to 79 cm^{-1} in the $\nu_1+\nu_3$ state and it vanishes in the $2\nu_2+\nu_4$ state.

Halonen et al [327] recorded the FT spectra of C_3O_2 in the 400–700 cm^{-1} region, analyzed the ν_5 and ν_6 band systems, and have reported the accurate wavenumber of the ν_6 fundamental band at 540.221 cm^{-1}. The ground state has $\sum_g^+$ symmetry and hence in absorption, the $\nu_5(\Pi_g)$ state is observed only in combination with excitation of odd quanta of the ν_7 (Π_u) bending mode. However, the $\nu_6(\Pi_u)$ state can be observed in absorption with the excitation of any number of quanta of ν_7. The effective ν_7 bending potential and associated manifold of energy levels in ν_5 and ν_6 were considered. The effective ν_7 bending potential in the ν_5 state showed a splitting, a "vibrational Renner–Teller effect", due to interaction between ν_5 and ν_7. Halonen et al [327] found that the rotational constant B_v for the $(\nu_5^1 + 2\nu_7^2)(\Pi_g)$ state showed some anomaly as the (observed)–(calculated) value was high, the reason for which, they could not explain.

Jensen [328], using the semi–rigid bender model, has analyzed the data available for the ν_5 state from Halonen et al [327]. An extended Hamiltonian describing the manifold of ν_7 bending mode superimposed on the ν_5 fundamental band was employed by Jensen [328], taking into account for the effects of vibrational angular momentum caused by the CCC bending motion (ν_7) and for *l*–doubling effects; and has reported the potential energy parameters for the ν_5 state.

Jensen and Johns [329] recorded the FT absorption spectrum of C_3O_2 in the 500–600 cm^{-1} region with a resolution of 0.004 cm^{-1}, and assigned the ν_6 fundamental band and some hot bands associated with ν_7,$2\nu_7$ and $3\nu_7$ states. A semi–rigid bender analysis yielded the effective CCC bending potential energy function ν_7 in the ν_6 state. Jensen and Johns [329] found that C_3O_2 is bent at equilibrium, with an equilibrium CCC bond angle of 156° and a barrier of 28 cm^{-1} to linearity exists. The anomalous rotational constant of the $(\nu_5^1 + 2\nu_7^2)(\Pi_g)$ level reported by Halonen et al [327] was explained as due to a $\Delta l = \pm 1$ Coriolis interaction with the $(\nu_6^1 + 3\nu_7^3)(\Delta_g)$ level [329].

Vander Auwera et al [330] have recorded the FT absorption spectrum of C_3O_2 between 500 and 600 cm^{-1} at a resolution of 0.002 cm^{-1}, assigned 23 bands corresponding to $\nu_6 + n\nu_7 \leftarrow n\nu_7$ with $n \leq 9$, and analyzed 20 bands. Even though the sample was cooled to –65°C, hot bands originating from levels with upto 8 quanta of ν_7 were observed. The data were fitted to the semi–rigid bender model to determine the CCC bending potential function in the ν_6 state.

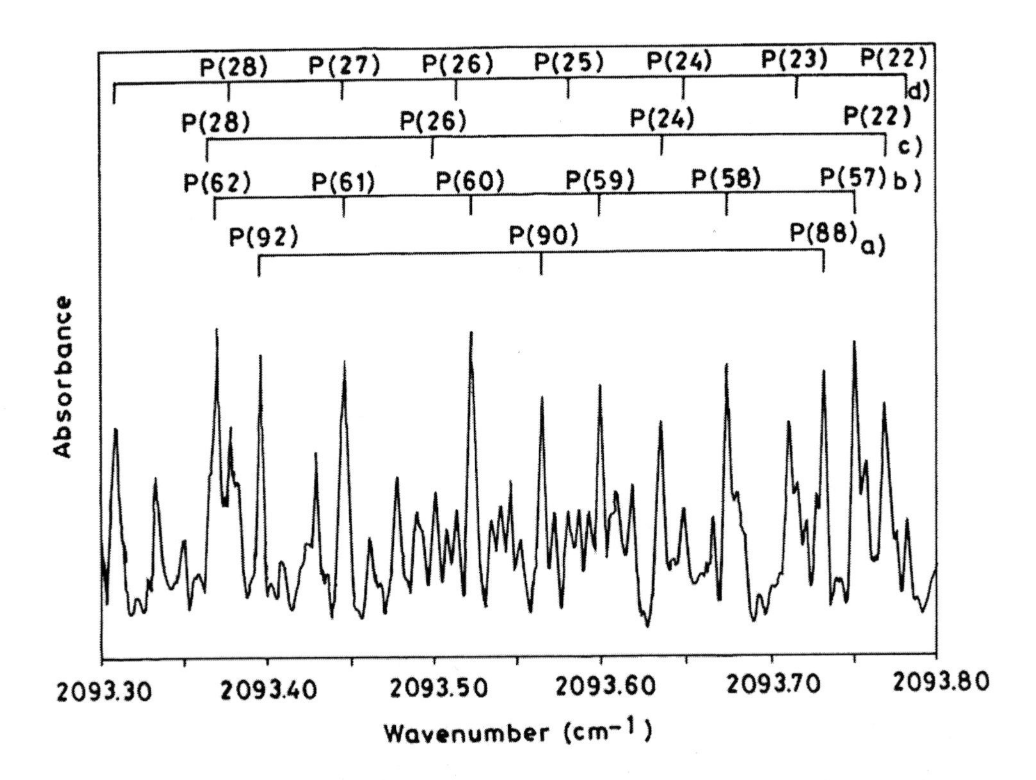

Figure 4.11. A part of the P–branch region of the ν_3 fundamental band in the FT absorption spectrum of SCCCS. The J assignments are a) ν_3 fundamental band, b) $\nu_3+\nu_7$ (Π_g) $\leftarrow\nu_7(\Pi_u)$ band, c) $\nu_3+2\nu_7$ $(\sum_u^+) \leftarrow 2\nu_7(\sum_g^+)$ subband, and d) $\nu_3+2\nu_7$ (Δ_u)$\leftarrow$ $2\nu_7$ (Δ_g) subband. (Reproduced from Holland et al [135] with permission from Elsevier Science & Academic Press, Orlando, Florida, USA).

Holland et al [135] have made high resolution measurements on the ν_3 band system of SCCCS in the 2100 cm^{-1} region with an FT spectrometer and a tunable diode laser spectrometer. A part of the P-branch region in the ν_3 fundamental band (FT spectrum) of SCCCS reported by Holland et al [135] is shown in figure 4.11. Since both ^{12}C and ^{32}S have nuclear spin I = 0, only the energy levels which are totally symmetric with respect to an interchange of identical nuclei are allowed. Hence only transitions (s$\leftarrow$s) for even J appear in the $\sum_u^+-\sum_g^+$ bands (e.g. ν_3 band) where the seperation between lines is 4B. This can be seen in fig. 4.11, where the P (88), P (90) and P(92) of the ν_3 band are shown. Similarly, the even P (J) lines are seen in the spectrum of the $\nu_3+2\nu_7(\sum_u^+) \leftarrow 2\nu_7(\sum_g^+)$ subband. In the hot band $\nu_3+\nu_7(\Pi_g)\leftarrow\nu_7(\Pi_u)$, since each rotational level in the Π vibrational state is split into two components with l_7-=1e and 1f, the transitions involving s$\leftarrow$s are observed, and both odd and even P(J) lines appear in the spectrum. Similar features are observed in the $\nu_3+2\nu_7$ $(\Delta_u)\leftarrow2\nu_7(\Delta_g)$ subband. The band centre of the ν_3 fundamental of SCCCS was determined to be 2100.098831 (35) cm^{-1} by Holland et al [135].

4.5 Six Atomic Molecules

In the literature, the reported work on the FT spectra of six atomic linear molecules are mainly on diacetylene and dicyanoacetylene.

(A) C_4H_2

Diacetylene (HC≡C–C≡CH) is a molecule of astrophysical interest, as it has been detected in the atmosphere of Titan. Absorption spectra of the molecule have been reported by some workers [333–335, 50] and references to earlier work are available in the papers of Guelachvili et al [333] and Khlifi et al [50]. Diacetylene has 13 vibrational modes, of which 9 fundamental vibrations are observed: five stretching modes ν_1 to ν_5 and four bending modes ν_6 to ν_9 (all degenerate). The frequencies of fundamental modes of vibration have been compiled by Guelachvili et al [333] and they are (in cm^{-1})

ν_1 $(\sum_g^+)$ = 3332.1541, ν_2 $(\sum_g^+)$ = 2188.9285, ν_3 $(\sum_g^+)$ = 871.9582, ν_4 $(\sum_u^+)$=3333.6647, ν_5 $(\sum_u^+)$ = 2022.2415, $\nu_6(\Pi_g)$ = 625.6436, $\nu_7(\Pi_g)$ = 482.7078, $\nu_8(\Pi_u)$ = 628.0409 and $\nu_9(\Pi_u)$ = 220.1236. Arie and Johns [335] have reported accurate values of ν_8 and ν_9 at 627.89423 (10) cm^{-1} and 219.97713 (10) cm^{-1} respectively. The schematic diagram of the normal vibrations of C_4H_2 are also available [ref. 111, p. 324].

Guelachvili et al [333] have recorded absorption spectra in the 1850–2523 and 2860–3584 cm^{-1} regions, analyzed ν_4 and ν_5 bands and their associated hot bands, several combination bands, and also have reported the accurate line positions and rotational constants for various vibrational levels of the ground state. The ν_4 band was found to be stronger than the ν_5 band. Dang–Nhu et al [334] have measured the absolute intensities of the P and R rotational lines of the ν_5 band in the 5μm region and obtained the value of band strength as S_0^v = 2.40±0.05 cm^{-1} atm^{-1} at 297K.

The bending energy levels ν_8 and ν_9, and hot bands in the 628 and 220 cm^{-1} regions have been investigated by Arie and Johns [335]. The states v_8=1, v_9=1 were found to be involved in vibrational and rotational l–resonance interactions and also in several perturbations. Using the notation $(v_8,v_9)^k$ with $k=|l_8+l_9|$, Arie and Johns [335] have suggested that Fermi interaction affects the $(1,1)_e^0$ levels.

Khlifi et al [50] have recorded the absorption spectra in the 250–4300 cm^{-1} region, have measured absolute band intensities for 11 bands, and have compared their values with the theoretical and experimental values reported by other workers, and found reasonable agreement with the earlier reported results. The possible contribution of C_4H_2 to the radiance spectrum of Titan has also been discussed by Khlifi et al [50].

(B) C_4N_2

The FT spectra of dicyanoacetylene (N≡C–C≡C–C≡N) have been recorded by Winther et al [336] in the 450–1200 cm^{-1} region at a resolution of 0.003 cm^{-1}. The bands ν_5, ν_8, combination bands $\nu_6+\nu_8$, $\nu_7+\nu_8$, $\nu_6+\nu_9$, and hot bands originating from $v_9=1$ (*e*, *f*) were analyzed and effective ro–vibrational constants for the states involved in the transitions reported. For the ground vibrational state, they reported accurate values of B_0 and D_0 as 0.4458699 (10) cm^{-1} and 1.049(4) x 10^{-9} cm^{-1} respectively.

The hot band systems $\nu_7+\nu_8+n\nu_9-n\nu_9$ (n=0 to 5), $\nu_6+(n+1)\nu_9-n\nu_9$ (n=0 to 7), and several bands originating from the $v_9=2$ level were analyzed and the molecular constants reported by Hegelund et al [337] who have also considered *l*–type resonance in the overtone level $v_9=2$, with $l_9=0,\pm2$. The $\nu_6+(n+1)\nu_9-n\nu_9$ band system was found to be strongly perturbed. In a later paper on the $\nu_6+n\nu_9$ levels of C_4N_2, Winther and Hegelund [134] analyzed some additional series of hot bands $\nu_6+(n+1)\nu_9-(n+2)\nu_9$ [n=0 to 6] with $\Delta l=0$ and $l=n+2$ near 397 cm^{-1}. For each of the level systems $\nu_6+\nu_9$, $\nu_6+2\nu_9$ and $\nu_6+3\nu_9$, they analyzed the upper state energies by setting up a Hamiltonian matrix incorporating both rotational and vibrational *l*–type resonances. The main reason for the strong perturbation of the $\nu_6+n\nu_9$ levels was attributed to *l*–type resonances. The effective anharmonicity and rotational constants for the $v_6=1$, $v_9=n$ (n=0 to 8, with *l* upto 9) have also been reported by Winther and Hegelund [134].

The wavenumber of the lowest stretching fundamental ν_3 has not been known precisely in gas phase. The $v_3=1$ level of C_4N_2 has Σ_g^+ symmetry and transition from the ground state (Σ_g^+) is forbidden. Quite recently, Winther et al [338] have recorded the high resolution FT spectrum of C_4N_2 in a sample cell with a path length of 172 m, in the 450–1700 cm^{-1} region, at resolutions between 0.0013 and 0.0018 cm^{-1}. They analyzed the $\nu_4-\nu_3$, $\nu_5-\nu_3$, $\nu_3-\nu_9$, $\nu_3+\nu_7+\nu_8-\nu_3$ and $\nu_3+\nu_7+\nu_8$ bands from which, a value for ν_3 = 606.3258 (3) cm^{-1} in gaseous phase has been reported [338].

4.6 Linear Molecules with More than Six Atoms

There are very few number of papers reported on the FT spectra of linear molecules with more than six atoms. A survey of the work on some molecules are presented in this subsection.

(A) Seven Atomic Molecules

The high resolution FTIR spectrum of the ν_4 band system of the cumulene molecule 1,2,3,4–pentatetraen–1,5–dione (OCCCCCO) in gaseous phase has been recorded by Holland et al [113] at an unapodized resolution of 0.022 cm^{-1}. The band centres for the $\nu_4(\Sigma_u^+)$, $\nu_5(\Sigma_u^+)$, $\nu_6(\Sigma_u^+)$, $\nu_9(\Pi_u)$ and $\nu_{10}(\Pi_u)$ bands were determined at 2242, 2065, 1152, 542 and 474 cm^{-1} respectively. The observed ro–vibrational data were analyzed with an expression involving powers of J(J+1) upto 5. Local resonances were observed at J′=87 and 119 of the ν_4 band.

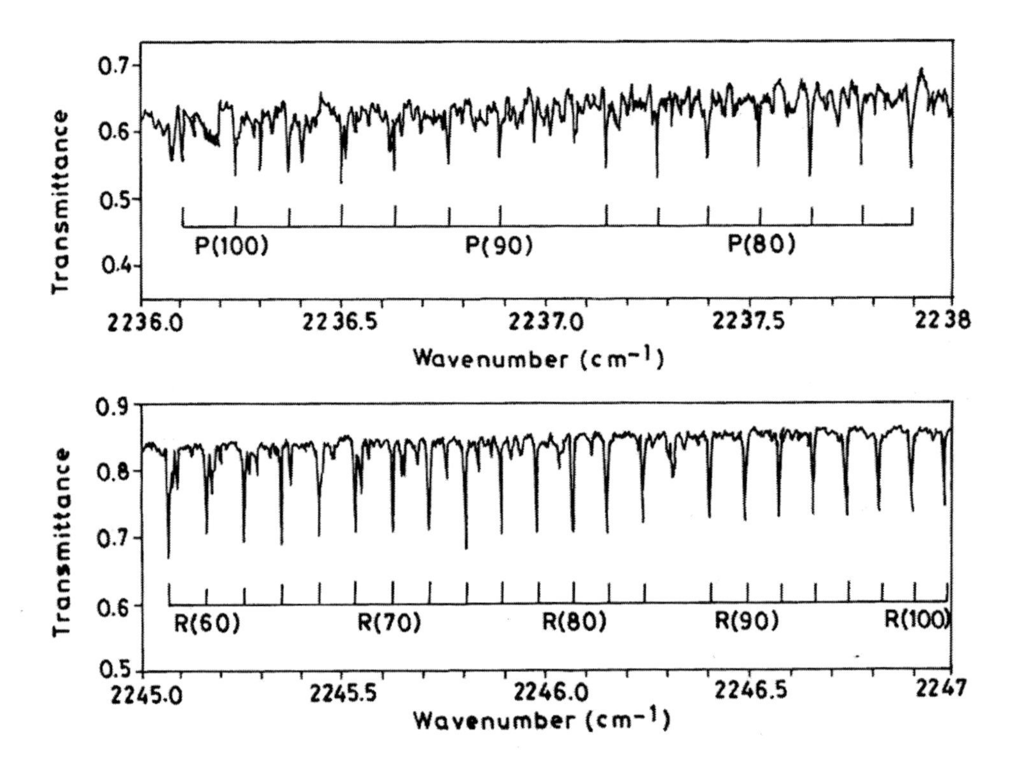

Figure 4.12. A part of the $\nu_4(\sum_u^+)$ band of OCCCCCO showing the P (upper trace) and R (lower trace) branch regions. The perturbation occurring at J′=87 can be seen in the traces in which the P(88) and R(86) lines are weak compared to the intensities of other rotational lines (Reproduced from Holland et al [113] with permission from Elsevier Science & Academic Press, Orlando, Florida, USA).

Figure 4.12 shows parts of the high resolution FTIR spectrum of C_5O_2 assigned to the ν_4 band at 2242 cm^{-1}. The P(J) and R(J) lines are indicated in the upper and lower traces respectively. The perturbation at J′=87 can be seen clearly in the traces in which the P(88) and R(86) lines are weak compared to the intensities of other rotational lines.

(B) Eight Atomic Molecules

The FT spectra of triacetylene (H–C≡C–C≡C–C≡C–H) have been reported by some authors [339–341, 44, 51]. Triacetylene exists in the dense atmosphere of Titan. The molecule has $D_{\infty h}$ symmetry, has no permanent electric dipole moment, and hence pure rotational transitions are forbidden. Information on the rotational and other constants are obtained from the analysis of vibration–rotation spectra. There are 13 normal modes of vibration: seven bond stretching vibrations ν_1 to ν_7 and six bond bending vibrations ν_8 to ν_{13} (doubly degenerate). Haas et al [51] have collected the calculated and observed band origins of the fundamental modes of vibrations of C_6H_2. Later, the more accurate wavenumbers for the band origins of ν_5[340] and ν_{11} [341] were reported. The observed wavenumbers (in cm^{-1}) for the triacetylene molecule are [51, 340, 341]: $\nu_1(\sum_g^+)$=3313(1), $\nu_2(\sum_g^+)$=2201(1),

$\nu_3(\Sigma_g^+)$=2019(1), $\nu_4(\Sigma_g^+)$=625(1), $\nu_5(\Sigma_u^+)$=3329.05164(49), $\nu_6(\Sigma_u^+)$= 2128.91637(32), $\nu_7(\Sigma_u^+)$=1115.0(5), $\nu_8(\Pi_g)$=622.38(40), $\nu_9(\Pi_g)$=491(1), $\nu_{10}(\Pi_g)$=258(1), $\nu_{11}(\Pi_u)$=621.340111(42), $\nu_{12}(\Pi_u)$=443.5(5) and $\nu_{13}(\Pi_u)$=105.038616(76) where the numbers in parentheses are the uncertainties in units of the last digit.

McNaughton and Bruget [339] have recorded the FT spectra of triacetylene in different ranges and studied the cold bands ν_5, ν_6, ν_{11}, $\nu_8+\nu_{11}$, and few hot bands originating from ν_{10}, ν_{11}, ν_{13} and $2\nu_{13}$ states. The $\nu_8+\nu_{11}$ $(\Sigma_u^+ - \Sigma_g^+)$ parallel band was found to be extensively perturbed, probably due to Coriolis resonance with some level. This band showed a local perturbation in the upper state centred around J′=89, in both P and R branches. The ν_6 band was also found to be strongly perturbed, while the ν_{11} band was unperturbed. The ν_{11} band showed an intense unresolved Q branch.

Matsumura et al [340] have made unique assignments for the ν_5 band of triacetylene from the FT spectra recorded at a resolution of 0.005 cm^{-1}. The odd J lines appeared stronger than the even J lines, due to the 3:1 nuclear statistical weights of the levels. Strong perturbations were observed in the ν_5 band at J′=20,50,54 and at other values. Fig. 4.13 shows the ν_5 band spectra around the P-branch (upper trace) and the R-branch (lower trace). In both branches, the P(J) and R(J) transitions with odd J values are observed as prominently strong lines appearing every 0.18 cm^{-1}. Due to perturbations at J′=50 and 54, the intensities of the P(51), P(55), R(49) and R(53) are weaker than those of the neighbouring transitions. Matsumura et al [340] have reported the accurate band origin of the ν_5 band and their reported value of rotational constant B_0=0.04417088(75) cm^{-1} agreed with the value reported by McNaughton and Bruget [339].

Haas et al [51] have recorded the spectrum of C_6H_2 in the region of the lowest fundamental band ν_{13} in the 27–227 cm^{-1} range with a resolution of 0.0018 cm^{-1}. The excited vibrational levels of ν_{13}=n were found to be populated to a noticeable degree at room temperature such that hot bands $(n+1)\nu_{13}-n\nu_{13}$ were observed upto n=3. The effective molecular parameters for ν_{13}=n (n=0,1,2,3) states and the *l*–doubling constant q_{13} of the ν_{13}=1 state were determined by Haas et al [51]. From the energies of the ν_{13}=1 and 2 levels, the quasilinearity parameter γ(defined by eq. 8 of ref. 51) was determined, which showed the C_6H_2 molecule to be a well behaved linear molecule.

Haas et al [341] have recorded the ν_{11} band of C_6H_2 near 622 cm^{-1} at a resolution of 0.0018 cm^{-1}, have given correct assignments for the fundamental ν_{11} band, and have revised the assignments for the $\nu_{11}+\nu_{13}\leftarrow\nu_{13}$ hot band. Some Q-branches were partially resolved and identified in the hot band. Intensity anomalies were observed in the same hot band which was explained due to the *l*–type resonance between Σ and Δ states of the $\nu_{11}=\nu_{13}$=1 state.

Delpech et al [44] have recorded the spectra of C_6H_2 in the 220–4300 cm^{-1} region, determined the absolute intensities of several bands, and have discussed the implications to Titan's atmosphere. The upper limit of the mean abundance of triacetylene in Titan's atmosphere has been estimated by Delpech et al [44] from the available IR spectra of Titan and from the laboratory measured absolute intensities of the ν_{11},ν_{12} and ν_9–ν_{13} bands of the molecule.

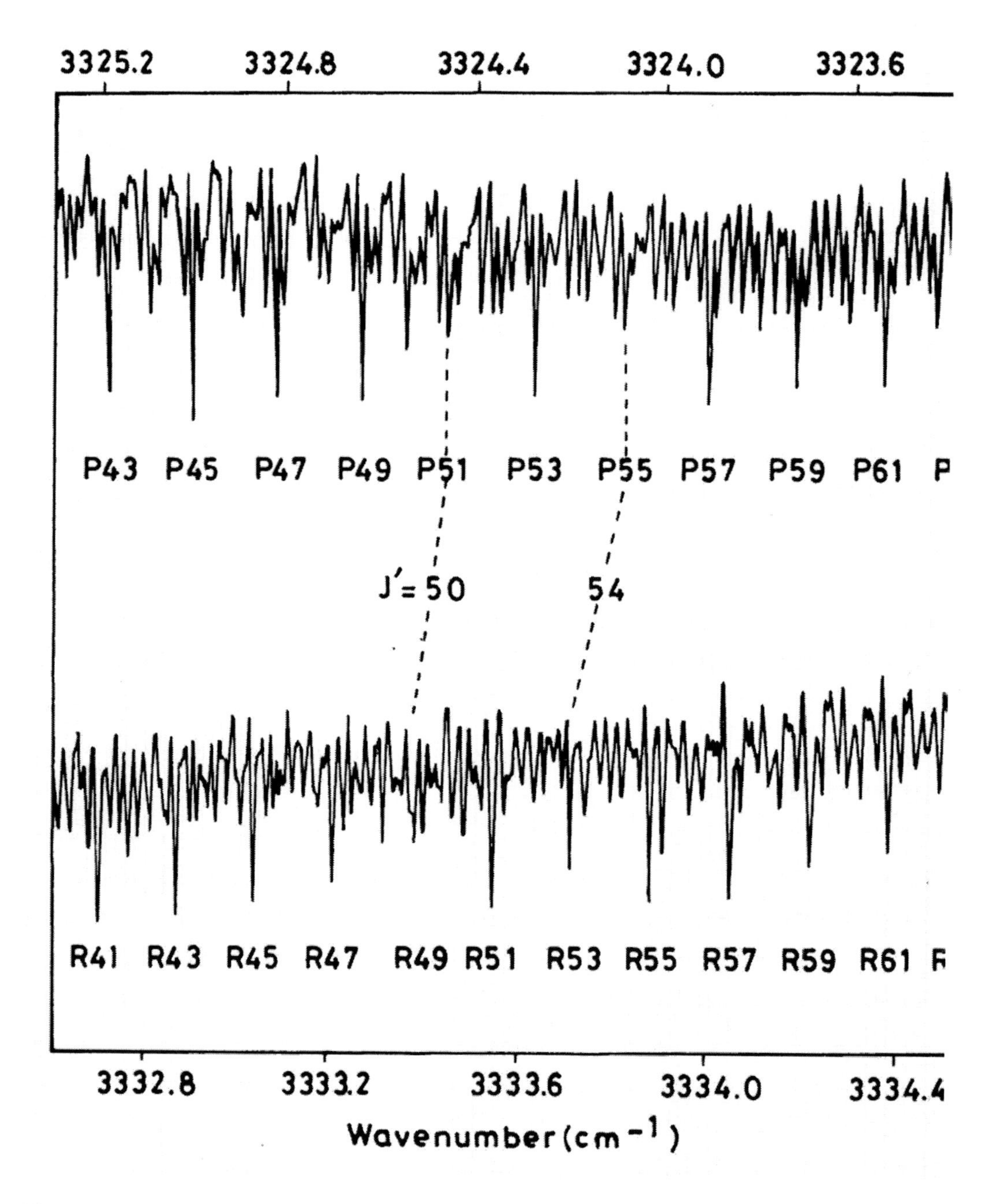

Figure 4.13. Th ν_5 band spectrum of triacetylene (HCCCCCCH) around the portions of the P branch (upper trace) and R branch (lower trace) regions. In both branches, due to the nuclear spin statistical weights of the levels, the lines with odd J values are observed as strong lines appearing at every 0.18 cm^{-1}. Due to perturbations which affected the J′=50 and 54 levels of the ν_5 state, the absorption intensities of P(51), P(55), R(49) and R(53) lines are weaker than those of the neighbouring transitions. (Reproduced from Matsumura et al [340] with permission from Elsevier Science & Academic Press, Orlando, Florida, USA).

(C) Ten Atomic Molecules

Recently, Shindo et al [43] have recorded the first FT spectrum of gaseous tetraacetylene (C_8H_2) in the 400–4000 cm^{-1} region. Due to the high instability of the molecule at room temperature, it was mixed with the solvent tetrabutyl tin and cooled to –50°C to trap the vapour. The spectra were obtained at a resolution of 2 cm^{-1} at 296K. The calculated values of the wavenumbers of the 17 vibrational modes of C_8H_2 by Shindo et al [table, 1, ref. 43] were useful in the identification of the bands observed.

Fig. 4.14 shows the FT spectrum of C_8H_2 in the range 400–4000 cm^{-1}, which shows three main bands assigned to ν_6, $\nu_{10}+\nu_{14}$ and ν_{14} modes and one very weak band assigned to the ν_8 mode. The spectrum of the solvent is shown as dotted lines. Shindo et al [43] used an optical path length of 10.6 m and a pressure of 1.8 x 10^{-2} mbar in their experiment.

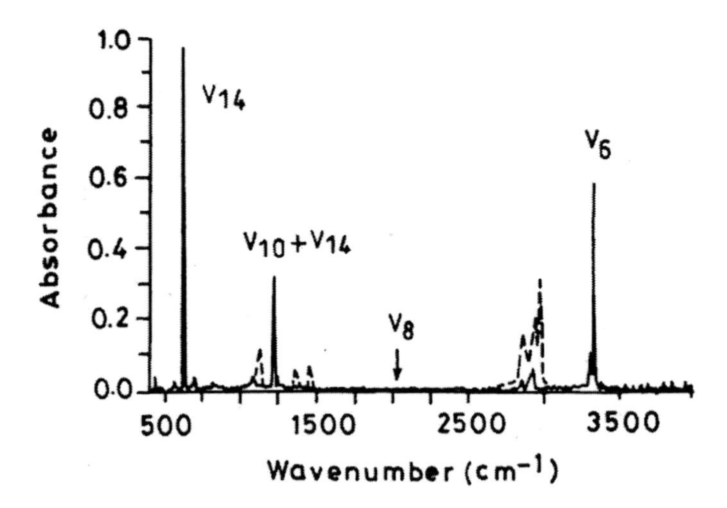

Figure 4.14. The FT spectrum (solid lines) of gaseous C_8H_2 in the range 400–4000 cm^{-1}, corrected from the solvent. The three main bands ν_6, $\nu_{10}+\nu_{14}$ and ν_{14} are indicated alongwith one weak ν_8 band. The spectrum was obtained at 293K at a resolution of 2 cm^{-1}, with an optical path length of 10.6m, and at a pressure of 1.8 x 10^{-2} mbar. The dotted line indicates the solvent (tetrabutyl tin) spectrum at a pressure of 5 x 10^{-3} mbar. (Reproduced from Shindo et al [43] with permission from Elsevier Science & Academic Press, Orlando, Florida, USA).

Fig. 4.15 shows the three main bands of C_8H_2 in some more detail. The vibration and transition symmetries involved are indicated for each band as well as the position of the P, Q and R branches. The band at 621 cm^{-1} is the ν_{14} band with a strong Q- branch due to the C≡C–H bending vibration. The band at 1220 cm^{-1} shows P-and R-branches due to combination $\nu_{10}+\nu_{14}$ (the combination of vibration symmetries $\pi_g\pi_u$ produces species $\sum_u^+ + \sum_u^- + \Delta_u$, where the $\sum_u^+$ species allows the $\sum - \sum$ transition, which shows only P and R branches). The third band at 3320 cm^{-1} is due to the ν_6 fundamental which also shows P and R branches only. From the spectra, Shindo et al [43] calculated the integrated band intensities for the three main bands, and have discussed the observational implications of the tetraacetylene bands in the atmosphere of Titan.

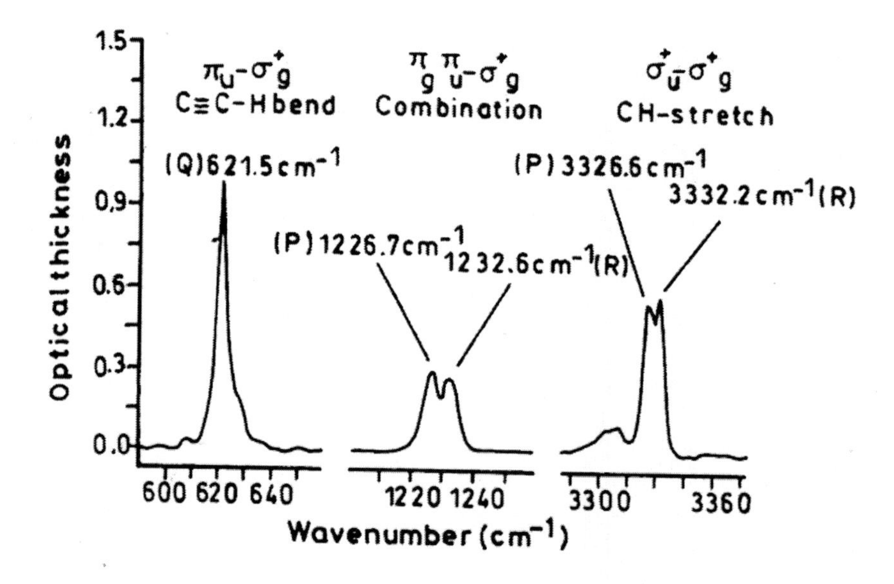

Figure 4.15. The three main bands of C_8H_2 (same as of fig 4.14) shown in more detail. The vibration and transition symmetries involved are indicated for each band as well as the positions of the P, Q and R branches. The features at 621 cm^{-1} (ν_{14} band), at 1220 cm^{-1} ($\nu_{10}+\nu_{14}$ combination band) and near 3320 cm^{-1} (ν_6 band) are shown alongwith the nature of vibrations involved (Reproduced from Shindo et al [43] with permission from Elsevier Science & Academic Press, Orlando, Florida, USA).

Chapter 5

CONCLUDING REMARKS

A great wealth of in–depth information on the energy levels, molecular dynamics and spectra of different types of linear and bent molecules have been obtained from high resolution FT spectroscopy. An attempt has been made in the present review to have an overview on the theoretical and experimental aspects involved in the interpretation of the high resolution FT spectra and also on the types of information obtained from linear molecules containing two to ten atoms. The description of the spectra, the perturbations observed, and the characteristic results obtained for several molecules, etc. have been included. However, only a small sample of the variety of methods of analysis of spectral data and typical results reported could be presented in this review. In fact, in the case of diatomic molecules alone, there are hundreds of papers reported, each molecule has been a case study by itself; presenting different types of spectral features, necessiating in different methods of analyses, and providing unique scientific information (a full review on diatomic molecules would have been apt).

The light linear molecules with large rotational constants have been ideal candidates to study the intricate aspects of molecular spectroscopy involving various anharmonic and *l*–tye interactions and resonances in their energy levels, some of which have been discussed in this review. The high quality of the spectra and spectral data, accurate calibration of the spectra, and nonlinear least squares fit to appropriate theoretical models (involving several parameters) resulting in low values of the standard deviation of the fit etc., are some of the advantageous features of high resolution FT spectroscopic research, which are highlighted with suitable examples in this review.

In the case of long chain linear molecules and heavy polyatomic molecules, in spite of the very high resolution of the FT spectrometers, it has not been possible to fully resolve the spectra. The spectral regions are densely crowded due to the effects of small rotational constants, presence of strong hot bands in the region, isotopic effects and due to several resonance effects. The vibrational and rotational *l*–type resonance effects become important at higher energies, as high density of interacting levels occur, leading to multiple channels of energy flow. It has been observed that the number of papers reported are rather low on linear molecules with more than four atoms. Attempts are required to unravel the mysteries of the unresolved strong Q-branches observed for many molecules, and to study the spectral features of long chain carbon molecules of astrophysical interest for e.g., of the polyynes $(H–(C{\equiv}C)_n–H)$. The spectral data from the longer polyynes and cyanopolyynes $(H–(C{\equiv}C)_n–N)$ are required to detect molecular signatures, to obtain information on abundances, and hence to provide parameters for the study of the interstellar medium and for planetary atmosphere modelling.

A number of research papers reporting high resolution emission spectra of diatomic molecules and absorption spectra of small and occasionally long linear molecules appear in the current literature. It is hoped that a timely review of the present type, is appropriate and would give a comprehensive glimpse on the importance and applications of high resolution Fourier transform spectroscopy in the modern molecular research scenario.

ACKNOWLEDGEMENTS

The author is very highly grateful to Elsevier Science and Academic Press, Orlando, Florida (USA) for kindly permitting me to reproduce the figures 4.1 to 4.15, the tables 4.1 to 4.7, and to use several theoretical formulas (many equations of section 3) and molecular constants of several molecules; from the published articles in the Journal of Molecular Spectroscopy, in this review article. The materials used are original to the text of Journal of Molecular Spectroscopy and full credit is given to JMS and Academic Press. Professor M. Herman (Bruxelles, Belgium) and American Institute of Physics, New York (USA) are thanked for permitting me to reproduce table 4.8 from the Journal of Chemical Physics. All the authors of the papers and books, whose contributions have been used/cited in this review, are thanked, because without the inclusion of some of the results of their published work, a review article of the present type would have been impossible.

The author is very much indebted to Professor G.Di Lonardo (Bologna, Italy) for kindly guiding me during my stay in Bologna for one year, without whose help, patience and perseverance, I would not have been able to understand the techniques and interpretations involved in high resolution FTIR spectroscopy. Encouragement from Professor Rajeshwari Prasad, Chairman, Department of Physics, Aligarh Muslim University, Aligarh, is also acknowledged.

REFERENCES

[1] R.J. Bell, *Introductory Fourier Transform Spectroscopy,* Academic Press, new York (1972).

[2] J. Chamberlain, *The Principles of Interferometric Spectroscopy,* John Wiley & Sons, New York (1979).

[3] P.R. Griffiths and J.A. De Haseth, *Fourier Transform Infrared Spectroscopy,* John Wiley & Sons, New York (1986).

[4] H.W. Schnopper and R.I. Thompson, "Fourier spectrometers" in *Methods of Experimental Physics* (edited by N. Carleton) **12A**, Academic Press, New York (1974).

[5] D. Oepts, "Fourier transform spectroscopy" in *Methods of Experimental Physics* (edited by D. Williams) **13B**, Academic Press, New York (1976).

[6] H. Sakai, "High resolving power Fourier spectroscopy" in *Spectrometric Techniques* (edited by G.A. Vanasse) Vol. I, Academic Press, New York (1977).

[7] G. Guelachvili, "Distortions in Fourier spectra and diagnosis" in *Spectrometric Techniques* (edited by G.A. Vanasse) Vol. II, Academic Press, New York (1981).

[8] J.R. Ferraro and L.J. Basile (editors) *Fourier Transform Infrared Spectroscopy Vol. I: Applications to Chemical Systems*, Academic Press, New York (1978).

[9] J.R. Ferraro and L.J. Basile (editors), *Fourier Transform Infrared Spectroscopy Vol. II: Applications to Chemical Systems,* Academic Press, Orlando, Florida (1979).

[10] J.R. Ferraro and L.J. Basile (editors), *Fourier Transform Infrared Spectroscopy vol. III: Techniques Using Fourier Transform Interferometry,* Academic Press, San Diego, California (1982).

[11] J.R. Ferraro and L.J. Basile (editors), *Fourier Transform Infrared Spectroscopy Vol. IV: Applications to Chemical Systems,* Academic Press, Orlando, Florida (1985).

[12] E. Fermi, *Z. Physik* **71**, 250 (1931).

[13] B. Darling and D. Dennison, *Phys. Rev* **57**, 128 (1940).

[14] A.G. Maki, G. C. Mellau, S. Klee, M. Winnewisser and W. Quapp, *J. Mol Spectrosc* **202**, 67 (2000).

[15] A.G. Maki and G.C. Mellau, *J. Mol. Spectrosc* **206**, 47 (2001).

[16] E. Mollmann, A.G. Maki, M. Winnewisser, B.P. Winnewisser and W. Quapp, *J. Mol. Spectrosc* **212**, 22 (2002).

[17] K.D. Setzer, C. Uibel, W. Zyrnicki, A. M. Pravilov, E.H. Fink, H.P. Liebermann, A.B. Alekseyev and R.J. Buenker, *J. Mol. Spectrosc* **204**, 163 (2000).

[18] K.D. Setzer, E.H. Fink, A.B. Alekseyev, H.P. Liebermann and R.J. Buenker, *J. Mol. Spectrosc* **206**, 181 (2001).

[19] K. Ziebarth, K.D. Setzer, O. Shestakov and E.H. Fink, *J. Mol. Spectrosc* **191**, 108 (1998).

[20] I. Dabrowski, D.W. Tokaryk, and J.K.G. Watson, *J. Mol. Spectrosc* **189**, 95 (1998).

[21] R.E. Gutterres, C.E. Fellows, J. Verges and C. Amiot, *J. Mol. Spectrosc* **206**, 62 (2001).

[22] Y. Krouti, A. Poclet, T. Hirao, B. Pinchemel and P.F. Bernath, *J. Mol. Spectrosc* **210**, 41 (2001).

[23] C. Focsa, H. Li and P.F. Bernath, *J. Mol. Spectrosc* **200**, 104 (2000).

[24] S.J. Park, S.W. Suh, Y.S. Lee and G.H. Jeung, *J. Mol. Spectrosc* **207**, 129 (2001).

[25] J. Vander Auwera, *J. Mol. Spectrosc* **201**, 143 (2000).

[26] J. L. Teffo, C.Claveau, Q. Kou, G. Guelachvili, A. Ubelmann, V.I. Perevalov and S. A. Tashkun, *J. Mol. Spectrosc* **201**, 249 (2000).

[27] L. Daumont, J. Vander Auwera, J.L. Teffo, V.I. Perevalov and S.A. Tashkun, *J. Mol. Spectrosc* **208**, 281 (2001).

[28] C. Chackerian. Jr, R. Freedman, L.P. Giver and L.R. Brown, *J. Mol. Spectrosc* **210**, 119 (2001).

[29] I.M. Grigoriev, R. Le Doucen, J. Boissoles, B. Calil, C. Boulet, J.M. Hartmann, X. Bruet and M.L. Dubernet, *J. Mol. Spectrosc* **198**, 249 (1999).

[30] L.R. Brown and C. Plymate, *J. Mol. Spectrosc* **199**, 166 (2000).

[31] J. Boissoles, F. Thibault, C. Boulet, J.P. Bouanich and J.M. Hartmann, *J. Mol. Spectrosc* **198**, 257 (1999).

[32] R.S. Pope and P.J. Wolf, *J. Mol. Spectrosc* **208**, 153 (2001).

[33] C. Chackerian Jr. R.S. Freedman, L.P. Giver and L.R. Brown, *J. Mol. Spectrosc* **192**, 215 (1998).

[34] (a).Y.I. Baranov and A.A. Vigasin, *J. Mol. Spectrosc* **193**, 319 (1999).
(b) A.A. Vigasin, Y.I. Baranov and G.V. Chlenova, *J. Mol. Spectrosc* **213**, 51 (2002).

[35] B. Mate, G.T. Fraser and W.J. Lafferty, *J. Mol. Spectrosc* **201**, 175 (2000).

[36] R.S. Ram and P.F. Bernath, *J. Mol. Spectrosc* **207**, 285 (2001).

[37] S. Civis, R. D'Cunha and K. Kawaguchi, *J. Mol. Spectrosc* **210**, 127 (2001).

[38] K.D. Setzer, E.H. Fink and D.A. Ramsay, *J. Mol. Spectrosc* **198**, 163 (1999).

[39] A. Jenouvrier, M.F. Merienne, B. Coquart, M. Carleer, S. Fally, A.C. Vandaele, C. Hermans and R. Colin, *J. Mol. Spectrosc* **198**, 136 (1999).

[40] M.F. Merienne, A. Jenouvrier, B. Coquart, S. Fally, R. Colin, A.C. Vandaele and C. Hermans, *J. Mol. Spectrosc* **202**, 171 (2000).

[41] S. Fally, A.C. Vandaele, M. Carleer, C. Hermans, A. Jenouvrier, M.F. Merienne, B. Coquart and R. Colin, *J. Mol. Spectrosc* **204**, 10 (2000).

[42] M.F. Merienne, A. Jenouvrier, B. Coquart, M. Carleer, S. Fally, R. Colin, A.C. Vandaele and C. Hermans, *J. Mol. Spectrosc* **207**, 120 (2001).

[43] F. Shindo, Y. Benilan, P. Chaquin, J.C. Guillemin, A. Jolly and F. Raulin, *J. Mol. Spectrosc* **210**, 191 (2001).

[44] C. Delpech, J.C. Guillemin, P. Paillous, M Khlifi, P. Bruston and F. Raulin, *Spectrochim Acta* **50A**, 1095 (1994).

[45] G.M. Plummer, G. Winnewisser, M. Winnewisser, J. Hahn and K. Reinartz, *J. Mol. Spectrosc* **126**, 255 (1987).

[46] Y. Kabbadj, M. Herman, G.Di Lonardo, L. Fusina and J. W.C. Johns, *J. Mol. Spectrosc* **150**, 535 (1991).
[47] G. Di Lonardo, P. Ferracuti, L. Fusina, E. Venuti and J.W.C. Johns, *J. Mol. Spectrosc* **161**, 466 (1993).
[48] A. Valentin, *Spectrochim. Acta* **51A**, 1127 (1995).
[49] J.J. Plateaux, A. Barbe and A. Delahaigue, *Spectrochim. Acta* **51A,** 1153 (1995).
[50] M. Khlifi, P. Paillous, C. Delpech, M. Nishio, P. Bruston and F. Raulin, *J. Mol. Spectrosc* **174**, 116 (1995).
[51] S. Haas, K.M.T. Yamada and G. Winnewisser, *J. Mol. Spectrosc* **164**, 445 (1994).
[52] H. Burger, S.Ma, J. Demaison, M. Le Guennec, C. Degli Esposti and L. Bizzochi, *J. Mol. Spectrosc* **199**, 109 (2000).
[53] D.B. Wang, T. Imajo, K. Tanaka, T. Tanaka and H. Burger, *J. Mol. Spectrosc* **207**, 70 (2001).
[54] J. U. White, *J. Opt. Soc. Am* **32**, 285 (1942).
[55] R. D'Cunha, Y. A. Sarma, G. Guelachvili, R. Ferrenq and K. Narahari Rao, *Indian J. Pure and Appl. Phys* **37,** 577 (1999).
[56] R.J. Kshirsagar, L.P. Giver and C. Chackerian Jr, *J. Mol. Spectrosc* **199**, 230 (2000).
[57] A. Farkhsi, H. Bredohl, I. Dubois, F. Remy and A. Fayt, *J. Mol. Spectrosc* **201**, 36 (2000).
[58] F. Ito, P. Klose, T. Takanaga, H. Takeo and H. Jones, *J. Mol. Spectrosc* **194**, 17 (1999).
[59] R. A. Toth, *J. Mol. Spectrosc* **197**, 158 (1999).
[60] T. Strugariu, S. Naim, A. Fayt, H. Bredohl, J.F. Blavier and I. Dubois, *J. Mol. Spectrosc* **189**, 206 (1998).
[61] C.E. Fellows, R.F. Gutterres, A.P.C Campos, J. Verges and C. Amiot, *J. Mol. Spectrosc* **197**, 19 (1999).
[62] C.E. Bosch, P. Crozet, A.J. Ross and J.M. Brown, *J. Mol. Spectrosc* **202**, 253 (2000).
[63] C. Amiot, *J. Mol. Spectrosc* **203**, 126 (2000).
[64] T. Hirao, C. Dufour, B. Pinchemel and P.F. Bernath, *J. Mol. Spectrosc* **202**, 53 (2000).
[65] A. Bernard, C. Effantin, E.A. Shenyavskaya and J. d'Incan, *J. Mol. Spectrosc* **207**, 211 (2001).
[66] E.G. Lee, J.Y. Seto, T. Hirao, P.F. Bernath and R.J. Le Roy, *J. Mol. Spectrosc* **194**, 197 (1999).
[67] O. Shestakov, R. Breidohr, H. Demes, K.D. Setzer and E.H. Fink, *J. Mol. Spectrosc* **190**, 28 (1998).
[68] H.Li, R. Skelton, C. Focsa, B. Pinchemel and P.F. Bernath, *J. Mol. Spectrosc* **203**, 188 (2000).
[69] R.F. Gutterres, J. Verges and C. Amiot, *J. Mol. Spectrosc* **200**, 253 (2000).
[70] R.S. Ram and P.F. Bernath, *J. Mol. Spectrosc* **195**, 299 (1999).
[71] M.D. Saksena and M.N. Deo, *J. Mol. Spectrosc* **208**, 64 (2001).
[72] R.S. Ram and P.F. Bernath, *J. Mol. Spectrosc* **203**, 9 (2000).
[73] R.S. Ram and P.F. Bernath, *J. Mol. Spectrosc* **201**, 267 (2000).
[74] R.S. Ram, P.F. Bernath and S.P. Davis, *J. Mol. Spectrosc* **210**, 110 (2001).
[75] T. Imajo, K. Tokieda, Y. Nakashima, K. Tanaka and T. Tanaka, *J. Mol. Spectrosc* **204**, 21 (2000).
[76] T. Imajo, D.B. Wang, K. Tanaka and T. Tanaka, *J. Mol. Spectrosc* **203**, 216 (2000).

[77] A. Campargue, D. Bailly, J.L. Teffo, S.A. Tashkun and V.I. Perevalov, *J. Mol. Spectrosc* **193**, 204 (1999).
[78] C. Focsa, P.F. Bernath, R. Mitzner and R. Colin, *J. Mol. Spectrosc* **192**, 348 (1998).
[79] T. Parekunnel, L. C. O'Brien, T.L. Kellerman, T. Hirao, M. Elhanine and P.F. Bernath, *J. Mol. Spectrosc* **206**, 27 (2001).
[80] D. Bailly and M. Vervloet, *J. Mol. Spectrosc* **209**, 207 (2001).
[81] D. Bailly, S.A. Tashkun, V.I. Perevalov, J.L. Teffo and Ph. Arcas, *J. Mol. Spectrosc* **197**, 114 (1999).
[82] L.C. O'Brien, M. Dulick and S.P. Davis, *J. Mol. Spectrosc* **195**, 328 (1999).
[83] G. Guelachvili and K. Narahari Rao, *Handbook of Infrared Standards*, Academic Press, Orlando, Florida (1986).
[84] G. Guelachvili, M. Birk, Ch.J. Borde, J.W. Brault, L.R. Brown, B. Carli, A.R.H. Cole, K.M. Evenson, A.Fayt, D. Hausamann, J.W.C. Johns, J. Kauppinen, Q. Kou, A.G. Maki, K. Narahari Rao, R.A. Toth, W. Urban, A. Valentin, J. Verges, G. Wagner, M.H. Wappelharst, J.S. Wells, B.P. Winnewisser and M. Winnewisser, *J. Mol. Spectrosc* **177**, 164 (1996) and *Spectrochim Acta* **52A**, 717 (1996).
[85] F. Alboni, G. Di Lonardo, P. Ferracuti, L. Fusina, E. Venuti and K.A. Mohamed, *J. Mol. Spectrosc* **169**, 148 (1995).
[86] J.M. Brown, J.T. Hougen, K.P. Huber, J.W.C. Johns, I.Kopp, H. Lefebvre–Brion, A.J. Merer, D.A. Ramsay, J. Rostas and R.N. Zare, *J. Mol. Spectrosc* **55**, 500 (1975).
[87] G. Herzberg, *Molecular Spectra and Molecular Structure I. Spectra of Diatomic Molecules*, Van Nostrand Reinhold company, New York. (1950).
[88] R.N. Zare, A.L. Schmeltekopf, D.L. Albritton and W.J. Harrop, *J. Mol. Spectrosc* **48**, 174 (1973).
[89] C. Amiot, J.P. Maillard and J. Chauville, *J. Mol. Spectrosc* **87**, 196 (1981).
[90] J.M. Brown, E.A. Coulbourne, J.K.G. Watson and F.D. Wayne, *J. Mol. Spectrosc* **74**, 294 (1979).
[91] M. Douay, S.A. Rogers and P.F. Bernath, *Mol. Phys.* **64**, 425 (1988).
[92] J.M. Brown and J.K. G. Watson, *J. Mol. Spectrosc* **65**, 65 (1977).
[93] J.A. Coxon, *J. Mol. Spectrosc* **58**, 1 (1975).
[94] R.F. Gutterres, J. Verges and C. Amiot, *J. Mol. Spectrosc* **196**, 29 (1999).
[95] A. Bernard, C. Effantin, J.d' Incan and J. Verges, *J. Mol. Spectrosc* **204**, 55 (2000).
[96] E.A. Shenyavskaya, M.A. Lebeault–Dorget, C. Effantin, J.d'Incan, A. Bernard and J. Verges. *J. Mol. Spectrosc* **171**, 309 (1995).
[97] S. Wallin, R. Koivisto and O. Launila, *J. Chem Phys* **105**, 388 (1996).
[98] J.L. Dunham, *Phys. Rev* **41**, 721 (1932).
[99] C.H. Townes and A.L. Schawlow. *Microwave Spectroscopy*, McGraw–Hill, London (1955) pp. 9–11.
[100] J.K.G. Watson, *J. Mol. Spectrosc* **80**, 411 (1980).
[101] P.F. Bernath, *Spectra of Atoms and Molecules*, Oxford University Press, New York (1995) p. 208.
[102] J.B. White, M. Dulick and P.F. Bernath, *J. Mol. Spectrosc* **169**, 410 (1995).
[103] A.G. Maki, Wm. Bruce Olson and G. Thompson, *J. Mol. Spectrosc* **144**, 257 (1990).
[104] M. Dulick, K.Q. Zhang, B. Guo and P.F. Bernath, *J. Mol. Spectrosc* **188**, 14 (1998).
[105] D. Bailly, C. Rossetti, F. Thibault and R. Le Doucen, *J. Mol. Spectrosc* **148**, 329 (1991).

[106] R. Herman and R.F. Wallis, *J. Chem. Phys* **23**, 637 (1955).
[107] C. Chackerian Jr, D. Goorvich and L.P. Giver, *J. Mol. Spectrosc* **113**, 373 (1985).
[108] K.M.T. Yamada, F.W. Birss and M.R. Aliev, *J. Mol. Spectrosc* **112**, 347 (1985).
[109] M. Niedenhoff and K.M.T Yamada, *J. Mol. Spectrosc* **157**, 182 (1993).
[110] J.K.G. Watson, *J. Mol. Spectrosc* **101**, 83 (1983).
[111] G. Herzberg, *Molecular Spectra and Molecular Structure II. Infrared and Raman Spectra of Polyatomic Molecules*, D. Van Nostrand Reinhold company, New York (1945).
[112] J.K. Holland, D.A. Newnham, I.M.Mills and M. Herman, *J. Mol. Spectrosc* **151**, 346 (1992).
[113] F. Holland, M. Winnewisser, G. Maier, H.P. Reisenauer and A. Ultrich, *J. Mol. Spectrosc* **130**, 470 (1988).
[114] M. Herman, T.R. Huet, Y. Kabbadj and J. Vander Auwera, *Mol. Phys* **72**, 75 (1991).
[115] H. Sarkkinen, *J. Mol. Spectrosc* **207**, 136 (2001).
[116] C. Vigouroux, A. Fayt, A. Guarnieri, A. Huckauf, H. Burger, D. Lentz and D. Preugschat, *J. Mol. Spectrosc* **202**, 1 (2000).
[117] T. Ahonen, S. Alanko, V.M. Horneman, M. Koivussari, R. Paso, A.M. Tolonen and R. Anttila, *J. Mol. Spectrosc* **181**, 279 (1997).
[118] G. Di Lonardo, P. Ferracuti, L. Fusina and E. Venuti, *J. Mol. Spectrosc* **164**, 219 (1994).
[119] D.M. Jonas, S.A.B. Solina, B. Rajaram, R.J. Silbey, R.W. Field, K. Yamanouchi and S. Tsuchiya, *J. Chem Phys* **99**, 7350 (1993).
[120] J. Pliva, *J. Mol. Spectrosc* **44**, 165 (1972).
[121] F. Winther, S. Klee, G. Mellau, S. Naim, L. Mbosei and A. Fayt, *J. Mol. Spectrosc* **175**, 354 (1996).
[122] B. Coveliers, W.K. Ahmed, A. Fayt and H. Burger, *J. Mol. Spectrosc* **156**, 77 (1992).
[123] M. Herman, M.I. El Idrissi, A. Pisarchik, A. Campargue, A.C. Gaillot, L. Biennier, G. Di Lonardo and L. Fusina, *J. Chem. Phys* **108**, 1377 (1998).
[124] I.M. Mills and A.G. Robiette, *Mol. Phys* **56**, 743 (1985).
[125] B.C. Smith and J. S. Winn, *J. Chem. Phys* **89**, 4638 (1988).
[126] O. Vaittinen, T. Lukka, L. Halonen, H. Burger and O. Polanz, *J. Mol. Spectrosc* **172**, 503 (1995).
[127] A. Maki, W. Quapp, S. Klee, G. Ch. Mellau and S. Albert, *J. Mol. Spectrosc* **180**, 323 (1996).
[128] O. Vaittinen, L. Halonen, H. Burger and O. Polanz, *J. Mol. Spectrosc* **167**, 55 (1994).
[129] G. Amat and H.H. Nielsen, *J. Mol. Spectrosc* **2**, 152 (1958).
[130] A.M. Tolonen and S. Alanko, *Mol. Phys* **75**, 1155 (1992).
[131] A.M. Ahonen, T. Ahonen and S. Alanko, *J. Mol. Spectrosc* **191**, 117 (1998).
[132] H. Sarkkinen, A.M. Ahonen and S. Alanko, *J. Mol. Spectrosc* **193**, 396 (1999).
[133] A.G. Maki and S. Klee, *J. Mol. Spectrosc* **193**, 183 (1999).
[134] F. Winther and F. Hegelund *J. Mol. Spectrosc* **189**, 270 (1998).
[135] F. Holland, M. Winnewisser, C. Jarman, H.W. Kroto and K.M.T. Yamada, *J. Mol. Spectrosc* **130**, 344 (1988).
[136] J.W. C. Johns and J. Vander Auwera, *J. Mol. Spectrosc* **140**, 71 (1990).
[137] J.W.C. Johns, Z. Lu, F. Thibault, R. Le Doucen, Ph. Arcas and Ch. Boulet, *J. Mol. Spectrosc* **159**, 259 (1993).

[138] J.K.G. Watson, *J. Mol. Spectrosc* **125**, 428 (1987).
[139] J.C. Grecu, B.P. Winnewisser and M. Winnewiser, *J. Mol. Spectrosc* **159**, 551 (1993).
[140] A. Faye, Q. Kou, R. Ferrenq and G. Guelachvili , *J. Mol. Spectrosc* **197**, 147 (1999).
[141] A.J. Ross, P. Crozet, C. Linton, F. Martin, I. Russier and A. Yiannopoulou, *J. Mol. Spectrosc* **191**, 28 (1998).
[142] K. Ziebarth, K.D. Setzer and E.H. Fink *J. Mol. Spectrosc* **173**, 488 (1995).
[143] E. H. Fink, K. D. Setzer, D.A. Ramsay, J.P. Towle and J.M. Brown, *J. Mol. Spectrosc* **178**, 143 (1996).
[144] E.H. Fink, H. Kruse, D.A. Ramsay and D.C. Wang, *Mol. Phys* **60**, 277 (1987).
[145] J.K.G. Watson, *Can. J. Phys* **46**, 1637 (1968).
[146] K.D. Setzer, J.B. Burnecka, W. Zyrnicki and E.H. Fink, *J. Mol. Spectrosc* **203**, 244 (2000).
[147] S. Rousseau, A.R. Allouche and M.A. Frecon, *J. Mol. Spectrosc* **203**, 235 (2000).
[148] R.S. Ram and P.F. Bernath, *J. Mol. Spectrosc* **122**, 275 (1987).
[149] T. Gustavsson, C. Amiot and J. Verges, *Mol Phys* **64**, 293 (1988).
[150] M. Douay, R. Nietmann and P.F. Bernath, *J. Mol. Spectrosc* **131**, 250 (1988).
[151] M. Douay, R. Nietmann and P.F. Bernath, *J. Mol. Spectrosc* **131**, 261 (1988).
[152] F. Roux and F. Michaud, *J. Mol. Spectrosc* **136**, 205 (1989).
[153] F. Roux and F. Michaud, *J. Mol. Spectrosc* **149**, 441 (1991).
[154] R.S. Ram, S. Tam and P.F. Bernath, *J. Mol. Spectrosc* **152**, 89 (1992).
[155] C.R. Brazier, *J. Mol. Spectrosc* **177**, 90 (1996).
[156] C. Amiot, M. Hafid and J. Verges, *J. Mol. Spectrosc* **180**, 121 (1996).
[157] F. Martin, R. Bacis, S. Churassy and J. Verges, *J. Mol. Spectrosc* **116**, 71 (1986).
[158] D. Cerny, R. Bacis and J. Verges, *J. Mol. Spectrosc* **116**, 458 (1986).
[159] T. Gustavsson, C. Amiot and J. Verges, *Mol. Phys* **64**, 279 (1988).
[160] C.E. Fellows, *J. Mol. Spectrosc* **136**, 369 (1989).
[161] M.A.L. Dorget, C. Effantin, A. Bernard, J.d'Incan, J. Chevaleyre and E.A. Shenyavskaya, *J. Mol. Spectrosc* **163**, 276 (1994).
[162] A.K. Adohi and S.G. Cotton, *J. Mol. Spectrosc* **190**, 171 ((1998).
[163] C. Amiot, J. Verges and C.R. vidal, *J. Mol. Spectrosc* **103**, 364 (1984).
[164] C. Amiot, *Mol. Phys* **58**, 667 (1986).
[165] C. Amiot and J. Verges, *Mol. Phys* **61**, 51 (1987).
[166] J.Y. Seto, Z. Morbi, F. Charron, S.K. Lee, P.F. Bernath and R.J. Le Roy, *J. Chem. Phys* **110**, 11756 (1999).
[167] H.G. Hedderich, M. Dulick and P.F. Bernath, *J. Chem. Phys* **99**, 8363 (1993).
[168] J.B. White, M.Dulick and P.F. Bernath, *J. Chem. Phys* **99**, 8371 (1993).
[169] J.M. Campbell, M. Dulick, D. Klapstein, J.B. White and P.F. Bernath, *J. Chem. Phys* **99**,8379 (1993).
[170] J. Ballard, W.B. Johnston, B.J. Kerridge and J.J. Remedios, *J. Mol. Spectrosc* **127**, 70 (1988).
[171] M.N. Spencer, C. Chackerian Jr., L.P. Giver and L.R. Brown, *J. Mol. Spectrosc* **165**, 506 (1994).
[172] M.N. Spencer, C. Chackerian Jr, L.P. Giver and L.R. Brown, *J. Mol. Spectrosc* **181**, 307 (1997).
[173] J.P. Bouanich, F. Rachet and A. Valentin, *J. Mol. Spectrosc* **178**, 157 (1996).
[174] S. Voigt, S. Dreher, J. Orphal and J. P. Burrows, *J. Mol. Spectrosc* **180**, 359 (1996).

[175] R.E. Thompson, J.H. Park, M.A.H. Smith, G.A. Harvey and J.M. Russell III, *J. Mol. Spectrosc* **106**, 251 (1984).
[176] S. Green and J. Hutson, *J. Chem. Phys* **100**, 891 (1994).
[177] K. Yoshino, J.R. Esmond, W.H. Parkinson, A.P. Thorne, R.C.M. Learner and G. Cox, *J. Chem. Phys* **111**, 2960 (1999).
[178] K. Yoshino, J. R. Esmond, W.H. Parkinson, A.P. Thorne, R.C.M. Learner, G. Cox and A.S-C. Cheung, *J. Chem Phys* **112**, 9791 (2000).
[179] V.P. Bellary and T.K. Balasubramanian, *J. Quant. Spectrosc. Radiat. Transfer* **45**, 283 (1991).
[180] K. Yoshino, J.R. Esmond, W.H. Parkinson, A.P. Thorne, J.E. Murray, R.C.M. Learner, G.Cox, A.S-C. Cheung, K.W.S. Leung, K. Ito, T. Matsui and T. Imajo, *J. Chem. Phys* **109**, 1751 (1998).
[181] T. Imajo, K. Yoshino, J.R. Esmond, W.H. Parkinson, A.P. Thorne, J. E. Murray, R. C.M. Learner, G. Cox, A.S-C. Cheung, K. Ito and T. Matsui, *J. Chem. Phys* **112**, 2251 (2000).
[182] J. Rufus, K. Yoshino, J.R. Esmond, A.P. Thorne, J.E. Murrray, T. Imajo, K. Ito and T. Matsui, *J. Chem. Phys* **115**, 3719 (2001).
[183] A. S-C. Cheung, D.H-Y. Lo, K.W-S. Leung, K. Yoshino, A.P. Thorne, J.E. Murray, K. Ito, T. Matsui and T. Imajo, *J. Chem. Phys* **116**, 155 (2002).
[184] G. Durry and G. Guelachvili, *J. Mol. Spectrosc* **168**, 82 (1994).
[185] J. Lindner, R.A. Loomis, J.J. Klaassen and S.R. Leone, *J. Chem. Phys* **108**, 1944 (1998).
[186] C.P. Rinsland, R. Zander, A. Goldman, F.J. Murcray, D. G. Murcray, M.R. Gunson and C.B. Farmer, *J. Mol. Spectrosc* **148**, 274 (1991).
[187] J.I. Choe, Y.M. Rho, S.M. Lee, A.C. Le Floch and S.G. Kukolich, *J. Mol. Spectrosc* **149**, 185 (1991).
[188] S.R. Langhoff and C.W. Bauschlicher Jr, *J. Mol. Spectrosc* **143**, 169 (1990).
[189] Y. Azuma, G. Huang, M.P.J. Lyne, A.J. Merer and V.I. Srdanov, *J. Chem. Phys* **100**, 4138 (1993).
[190] B.D. Rehfuss, M.H. Suh, T.A. Miller and V.E. Bondybey, *J. Mol. Spectrosc* **151**, 437 (1992).
[191] R.F. Gutterres, J. Verges and C. Amiot, *J. Mol. Spectrosc* **201**, 326 (2000).
[192] C.A. Leach, A.A. Tsekouras and R.N. Zare, *J. Mol. Spectrosc* **153**, 59 (1992).
[193] L.S. Rothman, C.P. Rinsland, A. Goldman, S.T. Massie, D.P. Edwards, J. M. Flaud, A. Perrin, C. Camy–Peyret, V. Dana, J.Y. Mandin, J. Schroeder, A.Mc Cann, R.R. Gamache, R.B. Wattson, K. Yoshino, K.V. Chance, K.W. Jucks, L.R. Brown, V. Nemtchinov and P. Varanasi, *J. Quant. Spectrosc. Radiat. Transfer* **60**, 665 (1998).
[194] K. Jolma, J. Kauppinen and V.M. Horneman, *J. Mol. Spectrosc* **101**, 300 (1983).
[195] K. Jolma, *J. Mol. Spectrosc* **111**, 211 (1985).
[196] M.P. Esplin and L.S. Rothman, *J. Mol. Spectrosc* **116**, 351 (1986).
[197] T. L. Tan, E.C.Looi and K.K. Lee, *J. Mol. Spectrosc* **157**, 261 (1993).
[198] C. Claveau, J.L. Teffo, D. Hurtmans and A. Valentin, *J. Mol. Spectrosc* **189**, 153 (1998).
[199] C. Claveau, J.L. Teffo, D. Hurtmans, A. Valentin and R. R. Gamache, *J. Mol. Spectrosc* **193**, 15 (1999).
[200] S.A. Tashkun, V.I. Pervalov and J.L. Teffo, *J. Mol. Spectrosc* **210**, 137 (2001).

[201] D. Bailly, R. Ferrenq, G. Guelachvili and C. Rossetti, *J. Mol. Spectrosc* **90**, 74 (1981).
[202] D. Bailly and C. Rossetti, *J. Mol. Spectrosc* **102**, 392 (1983).
[203] D. Bailly and C. Rossetti, *J. Mol. Spectrosc* **105**, 215 (1984).
[204] D. Bailly and C. Rossetti, *J. Mol. Spectrosc* **105**, 229 (1984).
[205] D. Bailly and C. Rossetti, *J. Mol. Spectrosc* **105**, 331 (1984).
[206] D. Bailly and C. Rossetti, *J. Mol. Spectrosc* **119**, 388 (1986).
[207] D. Bailly and N. Legay, *J. Mol. Spectrosc* **157**, 1 (1993).
[208] D. Bailly, *J. Mol. Spectrosc* **161**, 275 (1993).
[209] D. Bailly, C. Camy-Peyret and R. Lanquetin, *J. Mol. Spectrosc* **182**, 10 (1997).
[210] D. Bailly S.A. Tashkun, V.I. Perevalov, J. I. Teffo and Ph. Arcas, *J. Mol. Spectrosc* **190**, 1 (1998).
[211] D. Bailly, *J. Mol. Spectrosc* **192**, 257 (1998).
[212] D.C. Benner and C.P. Rinsland, *J. Mol. Spectrosc* **112**, 18 (1985).
[213] C.M. Deeley and J.W.C. Johns, *J. Mol. Spectrosc* **129**, 151 (1988).
[214] J.W.C. Johns, *J. Mol. Spectrosc* **134**, 433 (1989).
[215] C.B. Suarez and F.P.J. Valero, *J. Mol. Spectrosc* **140**, 407 (1990).
[216] L.P. Giver and C. Chackerian Jr, *J. Mol. Spectrosc* **148**, 80 (1991).
[217] J.W.C. Johns and M. Noel, *J. Mol. Spectrosc* **156**, 403 (1992).
[218] L.P. Giver, C. Chackerian Jr, M.N. Spencer, L.R. Brown and R.B. Wattson, *J. Mol. Spectrosc* **175**, 104 (1996).
[219] (a). S.A. Tashkun, V.I. Perevalov, J.L. Teffo, M. Lecoutre, T.R. Huet, A.Campargue, D. Bailly and M.P. Esplin, *J. Mol. Spectrosc* **200**, 162 (2000).
(b) J.L. Teffo, L. Daumont, C. Claveau, A. Valentin, S.A. Tashkun and V.I. Perevalov, *J. Mol. Spectrosc* **213**, 145 (2002).
[220] J.W.C. Johns, *J. Mol. Spectrosc* **125**, 442 (1987).
[221] M.M-Maclou, P. Dahoo, A. Henry, A. Valentin and L. Henry, *J. Mol. Spectrosc* **131**, 21 (1988).
[222] J.Y. Mandin, V. Dana, M.Y. Allout, L. Regalia, A. Barbe and J.J. Plateaux, *J. Mol. Spectrosc* **170**, 604 (1995).
[223] J.Y. Mandin, V. Dana, M. Badaoui, G. Guelachvili, M.M-Chapey, Q. Kou, R.B. Wattson and L.S. Rothman, *J. Mol. Spectrosc* **155**, 393 (1992).
[224] Q. Kou and G. Guelachvili, *J. Mol. Spectrosc* **148**, 324 (1991).
[225] R. Le Doucen and C. Boulet, *Spectrochim Acta* **51A**, 1239 (1995).
[226] M.M-Maclou, F. Rachet, C. Boulet, A. Henry and A. Valentin, *J. Mol. Spectrosc* **172**, 1 (1995).
[227] W.D. Gillespie, C.J. Meinrenken, W.R. Lempert and R.B. Miles, *J. Chem. Phys* **107**, 5995 (1997).
[228] F. Rachet, M.M-Maclou, A. Henry and A. Valentin, *J. Mol. Spectrosc* **175**, 315 (1996).
[229] J.I. Choe, T.Tipton and S.G. Kukolich, *J. Mol. Spectrosc* **117**, 292 (1986).
[230] J.I. Choe, D.K. Kwak and S.G. Kukolich, *J. Mol. Spectrosc* **121**, 75 (1987).
[231] A.H. Smith, S.L. Coy, W. Klemperer and K.K. Lehmann, *J. Mol. Spectrosc* **134**, 134 (1989).
[232] G. Duxbury and Y. Gang, *J. Mol. Spectrosc* **138**, 541 (1989).
[233] J. Hietanen, K. Jolma and V.M. Horneman, *J. Mol. Spectrosc* **127**, 272 (1988).
[234] W. Quapp, S. Klee, G. Ch. Mellau, S. Albert and A. Maki, *J. Mol. Spectrosc* **167**, 375 (1994).

[235] W. Quapp, V. Melnikov and G. Ch. Mellau, *J. Mol. Spectrosc* **211**, 189 (2002).
[236] W. Quapp, M. Hirsch, G. Ch. Mellau, S. Klee, M. Winnewisser and A. Maki, *J. Mol. Spectrosc* **195**, 284 (1999).
[237] G. Weirauch, A.A. Kachanov, A. Campargue, M. Bach, M. Herman and J. Vander Auwera, *J. Mol. Spectrosc* **202**, 98 (2000).
[238] M. El Azizi, F. Rachet, A. Henry, M.M-Maclou and A. Valentin, *J. Mol. Spectrosc* **164**, 180 (1994).
[239] F. Rachet, M.M-Maclou, M. El Azizi, A. Henry and A. Valentin *J. Mol. Spectrosc* **164**, 196 (1994).
[240] F. Rachet, M.M-Maclou, M. El Azizi, A. Henry and A. Valentin, *J. Mol. Spectrosc* **166**, 79 (1994).
[241] J.W.C. Johns, Z. Lu, M. Weber, J.M. Sirota and D.C. Reuter, *J. Mol. Spectrosc* **177**, 203 (1996).
[242] M. Weber, J.M. Sirota and D.C. Reuter, *J. Mol. Spectrosc* **177**, 211 (1996).
[243] J.M. Hartmann, J.P. Bouanich, K.W. Jucks, Gh. Blaquet, J.Walrand, D.Bermejo, J.L. Domenech and N. Lacome, *J. Chem. Phys* **110**, 1959 (1999).
[244] N. Hunt, S.C. Foster, J.W. C. Johns and A.R.W. McKellar, *J. Mol. Spectrosc* **111**, 42 (`1985).
[245] K. Jolma, V.M. Horneman, J. Kauppinen and A.G. Maki, *J. Mol. Spectrosc* **113**, 167 (1985).
[246] J.S. Wells, M.D. Vanek and A.G. Maki, *J. Mol. Spectrosc* **135**, 84 (1989).
[247] A.M. Tolonen, V.M. Horneman and S. Alanko, *J. Mol. Spectrosc* **144**, 18 (1990).
[248] A.G. Maki, J.S. Wells and J.B. Burkholder, *J. Mol. Spectrosc* **147**, 173 (1991).
[249] T.L. Tan and E.C. Looi, *J. Mol. Spectrosc* **148**, 262 (1991).
[250] T.L. Tan, E.C. Looi and K.T. Lua, *J. Mol. Spectrosc* **148**, 265 (1991).
[251] V.M. Horneman, M.Koivussari, A.M. Tolonen, S. Alanko, R. Anttila, R. Paso and T. Ahonen, *J. Mol. Spectrosc* **155**, 298 (1992).
[252] A. Belafhal, A. Fayt and G. Guelachvili, *J. Mol. Spectrosc* **174**, 1 (1995).
[253] E. Rbaihi, A. Belafhal, J. Vander Auwera, S. Naim and A. Fayt, *J. Mol. Spectrosc* **191**, 32 (1998).
[254] S. Naim, A. Fayt, H. Bredohl, J.F. Blavier and I. Dubois, *J. Mol. Spectrosc* **192**, 91 (1998).
[255] A. Fayt, R. Vandenhaute and J.G. Lahaye, *J. Mol. Spectrosc* **119**, 233 (1986).
[256] Q. Errera, J. Vander Auwera, A. Belafhal and A. Fayt, *J. Mol. Spectrosc* **173**, 347 (1995).
[257] H. Suguro, T. Konno, K. Sueoka, Y. Hamada and H. Uehara, *J. Mol. Spectrosc* **124**, 46 (1987).
[258] K. Sueoka, Y. Hamada and H. Uehara, *J. Mol. Spectrosc* **127**, 370 (1988).
[259] H. Burger, M. Litz, H. Willner, M. Le Guennec, G. Wlodarczak and J. Demaison, *J. Mol. Spectrosc* **146**, 220 (1991).
[260] M. Le Guennec, G. Wlodarczak, J. Demaison, H. Burger, M. Litz and H. Willner, *J. Mol. Spectrosc* **157**, 419 (1993).
[261] M. Litz, H. Burger, L.S. Masukidi, A. Fayt, J. Cosleou, P. Drean, L. Margules and J. Demaison, *J. Mol. Spectrosc* **196**, 155 (1999).
[262] C-L.C. Cheng, J.L. Hardwick and T.R. Dyke, *J. Mol. Spectrosc* **179**, 205 (1996).

[263] G. Blanquet, J. Walrand, J.F. Blavier, H. Bredohl and I. Dubois, *J. Mol. Spectrosc* **152**, 137 (1992).

[264] J. Walrand, G. Blanquet, J. Gustin, J.F. Blavier, H. Bredohl and I. Dubois, *J. Mol. Spectrosc* **174**, 85 (1995).

[265] G. Blanquet, J. Walrand, H. Bredohl and I. Dubois, *J. Mol. Spectrosc* **198**, 43 (1999).

[266] A. Farkhsi, H. Bredohl, I. Dubois and F. Remy, *J. Mol. Spectrosc* **181**, 119 (1997).

[267] F. Meyer, C. Meyer, H. Bredohl, I. Dubois, A. Saouli and G. Blanquet, *J. Mol. Spectrosc* **158**, 247 (1993).

[268] A. Saouli, I. Dubois, J.F. Blavier, H. Bredohl, G. Blanquet, C. Meyer and F. Meyer, *J. Mol. Spectrosc* **165**, 349 (1994).

[269] A. Saouli, I. Dubois and H. Bredohl, *J. Mol. Spectrosc* **170**, 228 (1995).

[270] A. Saouli, H. Bredohl, I. Dubois and A. Fayt, *J. Mol. Spectrosc* **174**, 20 (1995).

[271] F. Tamassia, C.D. Esposti, L. Bizzocchi, Z. Zelinger, M. Le Guennec, J. Demaison, M. Litz and H. Burger, *J. Mol. Spectrosc* **189**, 264 (1998).

[272] G. Thompson and A. G. Maki, *J. Mol. Spectrosc* **160**, 73 (1993).

[273] J. Lavigne, C. Pepin and A. Cabana, *J. Mol. Spectrosc* **104**, 49 (1984).

[274] C. Pepin and A. Cabana, *J. Mol. Spectrosc* **119**, 101 (1986).

[275] D.McNaughton and D.N. Bruget, *J. Mol. Spectrosc* **161**, 336 (1993).

[276] D.S. Han, C.M.L. Rittby and W.R.M. Graham, *J. Chem Phys* **108**, 3504 (1998).

[277] R. Cireasa, D. Cossart, M. Vervloet and J.M. Robbe, *J. Chem. Phys* **112**, 10806 (2000).

[278] V.M. Horneman, S. Alanko and J. Hietanen, *J. Mol. Spectrosc* **135**, 191 (1989).

[279] J.J. Hillman, D.E. Jennings, G.W. Halsey, S. Nadler and W.E. Blass, *J. Mol. Spectrosc* **146**, 389 (1991).

[280] R.D'Cunha, Y.A. Sarma, G. Guelachvili, R. Ferrenq, Q. Kou, V.M. Devi, D.C. Benner and K.N. Rao, *J. Mol. Spectrosc* **148**, 213 (1991).

[281] J. Vander Auwera, D. Hurtmans, M. Carleer and M. Herman, *J. Mol. Spectrosc* **157**, 337 (1993).

[282] R.D'Cunha, Y.A. Sarma, V.A. Job, G. Guelachvili and K.N. Rao, *J. Mol. Spectrosc* **157**, 358 (1993).

[283] Y.A. Sarma, R.D'Cunha, G. Guelachvili, R. Ferrenq and K.N. Rao, *J. Mol. Spectrosc* **173**, 561 (1995).

[284] J. Hietanen, V.M. Horneman and J. Kauppinen, *Mol. Phys* **59**, 587 (1986).

[285] K.A. Mohamed, *Indian. J. Pure. Appl. Phys* **38**, 681 (2000).

[286] K.A. Mohamed, *Indian. J. Pure. Appl. Phys* (2002) in press.

[287] K.A. Mohamed, *Indian. J. Pure. Appl. Phys* **36**, 705 (1998).

[288] K.A. Mohamed, *Indian. J. Pure. Appl. Phys* **35**, 402 (1997).

[289] R.D'Cunha, Y.A. Sarma, G. Guelachvili, R. Ferrenq and K.N. Rao, *J. Mol. Spectrosc* **160**, 181 (1993).

[290] K.A. Keppler, G. Ch. Mellau, S. Klee, B.P. Winnewisser, M. Winnewisser, J. Pliva and K.N. Rao, *J. Mol. Spectrosc* **175**, 411 (1996).

[291] W.J. Lafferty and A.S. Pine, *J. Mol. Spectrosc* **141**, 223 (1990).

[292] E. Venuti, G.Di Lonardo, P. Ferracuti, L. Fusina and I.M. Mills, *Chem. Phys* **190**, 279 (1995).

[293] J. Lievin, M.A. Temsamani, P. Gaspard and M. Herman, *Chem. Phys* **190**, 419 (1995).

[294] M. Weber W.E. Blass, G.W. Halsey and J.J. Hillman, *J. Mol. Spectrosc* **165**, 107 (1994).

[295] G.Di Lonardo, A Baldan, G. Bramati and L. Fusina, *J. Mol. Spectrosc* **213**, 57 (2002).
[296] M. Herman, D. Hurtmans, J. Vander Auwera and M. Vervloet, *J. Mol. Spectrosc* **150**, 293 (1991).
[297] M.A. Temsamani and M. Herman, *J. Chem. Phys* **102**, 6371 (1995).
[298] M.A. Temsamani and M. Herman, *J. Chem. Phys* **105**, 1355 (1996).
[299] M.A. Temsamani, M. Herman, S. A.B. Solina, J.P. O'Brien and R.W. Field, *J. Chem. Phys* **105**, 11357 (1996).
[300] M.I. El Idrissi, J. Lievin, A. Campargue and M. Herman, *J. Chem. Phys* **110**, 2074 (1999).
[301] G.Di Lonardo, L. Fusina, E. Venuti, J.W.C. Johns, M.I. El Idrissi, J. Lievin and M. Herman, *J. Chem. Phys* **111**, 1008 (1999).
[302] B.I. Zhilinskii, M.I. El Idrissi and M. Herman, *J. Chem. Phys* **113**, 7885 (2000).
[303] J.K. Holland, W.D. Lawrence and I.M. Mills, *J. Mol. Spectrosc* **151**, 369 (1992).
[304] A.F. Borrow, I.M. Mills and A. Mose, *Chem. Phys* **190**, 363 (1995).
[305] D.McNaughton and M. Shallard, *J. Mol. Spectrosc* **165**, 185 (1994).
[306] M. Halonen, *J. Mol. Spectrosc* **167**, 225 (1994).
[307] R. Brotherus, O. Vaittinen, L. Halonen, H. Burger and O. Polanz, *J. Mol. Spectrosc* **193**, 137 (1999).
[308] A.M. Tolonen, S. Alanko, M. Koivusaari, R. Paso and V.M. Horneman, *J. Mol. Spectrosc* **165**, 249 (1994).
[309] J. C. Grecu, B.P. Winnewisser and M. Winnewisser, *J. Mol. Spectrosc* **159**, 534 (1993).
[310] W. Quapp, A. Maki, S. Klee and G. Mellau, *J. Mol. Spectrosc* **187**, 126 (1998).
[311] M. Hochlaf, *J. Mol. Spectrosc* **207**, 269 (2001).
[312] J.W.G. Siebert, M. Winnewisser and B.P. Winnewisser, *J. Mol. Spectrosc* **180**, 26 (1996).
[313] F. Stroh, M. Winnewisser and B.P. Winnewisser, *J. Mol. Spectrosc* **162**, 435 (1993).
[314] H. Burger and S. Sommer, *J. Mol. Spectrosc* **151**, 148 (1992).
[315] H. Burger, M. Senzlober and S. Sommer, *J. Mol. Spectrosc* **164**, 570 (1994).
[316] H. Burger, S.Sommer, D. Lentz and D. Preugschat, *J. Mol. Spectrosc* **156**, 360 (1992).
[317] M. Khlifi, F. Raulin, E. Arie and G. Graner, *J. Mol. Spectrosc* **143**, 209 (1990).
[318] E. Arie, M. Dang Nhu, Ph. Arcas, G. Graner, H. Burger, G. Pawelke, M. Khlifi and F. Raulin, *J. Mol. Spectrosc* **143**, 318 (1990).
[319] K.M.T. Yamada and R.A. Creswell, *J. Mol. Spectrosc* **116**, 384 (1986).
[320] G.M. Plummer, D. Mauer, K.M.T. Yamada and K. Moller, *J. Mol. Spectrosc* **130**, 407 (1988).
[321] M. Khlifi, F. Raulin and M. Dang Nhu, *J. Mol. Spectrosc* **155**, 77 (1992).
[322] A. W. Manz, P. Connes, G. Guelachvili and C. Amiot, *J. Mol. Spectrosc* **54**, 43 (1975).
[323] W.H. Weber, P.D. Maker and C.W. Peters, *J. Chem. Phys* **64**, 2149 (1976).
[324] C.W. Peters, W.H. Weber and P.D. Maker, *J. Mol. Spectrosc* **66**, 133 (1977).
[325] W.H. Weber, P.D. Maker and C.W. Peters, *J. Mol. Spectrosc* **77**, 139 (1979).
[326] L. Fusina, I.M. Mills and G. Guelachvili, *J. Mol. Spectrosc* **79**, 101 (1980).
[327] L. Halonen, I M. Mills and J. Kauppinen, *Mol. Phys* **43**, 913 (1981).
[328] P. Jensen, *J. Mol. Spectrosc* **104**, 59 (1984).
[329] P. Jensen and J.W.C. Johns, *J. Mol. Spectrosc* **118**, 248 (1986).
[330] J. Vander Auwera, J.K. Holland, P. Jensen and J.W.C. Johns, *J. Mol. Spectrosc* **163**, 529 (1994).

[331] L. Fusina and I.M. Mills, *J. Mol. Spectrosc* **79**, 123 (1980).

[332] W.H. Weber, *J. Mol. Spectrosc* **79**, 396 (1980).

[333] G. Guelachvili, A.M. Craig and D.A. Ramsay, *J. Mol. Spectrosc* **105**, 156 (1984).

[334] M. Dang Nhu, G. Guelachvili and D.A. Ramsay, *J. Mol. Spectrosc* **146**, 513 (1991).

[335] E. Arie and J.W.C. Johns, *J. Mol. Spectrosc* **155**, 195 (1992).

[336] F. Winther, M. Schonhoff, R. Le Prince, A. Guarnieri, D.N. Bruget and D.McNaughton, *J. Mol. Spectrosc* **152**, 205 (1992).

[337] F. Hegelund, H. Mader, M. Wiemeler and F. Winther, *J. Mol. Spectrosc* **171**, 22 (1995).

[338] F. Winther, C. Vigouroux, A. Fayt, V.M. Horneman and R. Anttila, *J. Mol. Spectrosc* **212**, 223 (2002).

[339] D.McNaughton and D.N. Bruget, *J. Mol. Spectrosc* **150**, 620 (1991).

[340] K. Matsumura, K. Kawaguchi, D.McNaughton and D.N. Bruget, *J. Mol. Spectrosc* **158**, 489 (1993).

[341] S. Haas, G. Winnewisser, K.M.T. Yamada, K. Matsumura and K. Kawaguchi *J. Mol. Spectrosc* **167**, 176 (1994).

INDEX

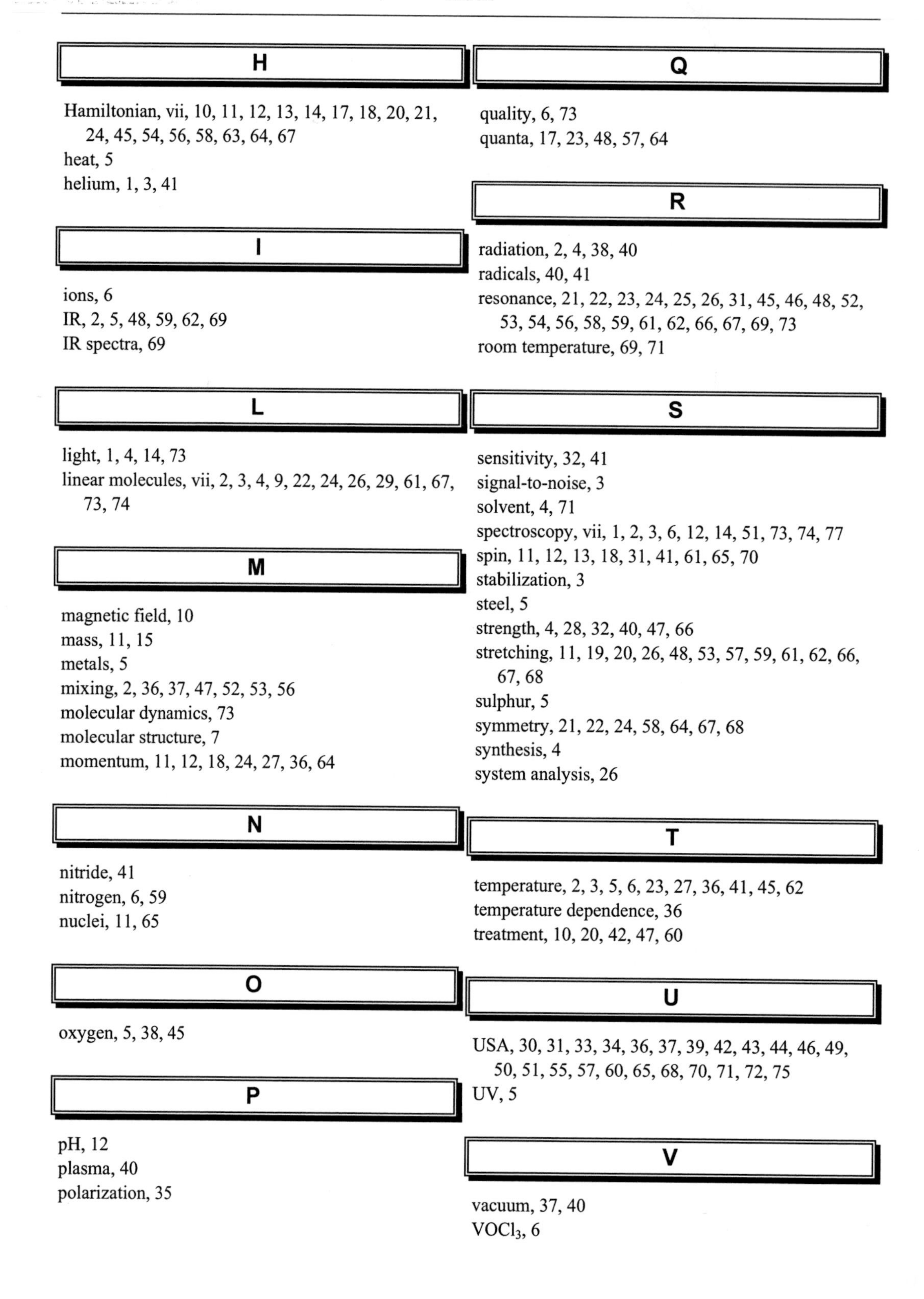

H

I

L

M

N

O

P

Q

R

S

T

U

V